HIGH ENERGY DENSITY TECHNOLOGIES
IN MATERIALS SCIENCE

HIGH ENERGY DENSITY TECHNOLOGIES IN MATERIALS SCIENCE

Proceedings of the 2nd IGD Scientific Workshop,
Novara, May 3–4, 1988

Edited by

F. GARBASSI and E. OCCHIELLO

Istituto Guido Donegani, Novara, Italy

KLUWER ACADEMIC PUBLISHERS

DORDRECHT / BOSTON / LONDON

Library of Congress Cataloging in Publication Data

IGD Scientific Workshop (2nd : 1988 : Novara, Italy)
   High energy density technologies in materials science :
proceedings of the 2nd IGD Scientific Workshop, Novara, May 3-4,
1988 / edited by F. Garbassi and E. Occhiello.
       p.   cm.

   1. Lasers--Industrial applications--Congresses. 2. Materials-
-Effect of radiation on--Congresses.   I. Garbassi, F.
II. Occhiello, E.   III. Title.
TA1673.I37 1988
621.36'6--dc20   89-24639

ISBN-13: 978-94-010-6710-2   e-ISBN-13: 978-94-009-0499-6
DOI: 10.1007/978-94-009-0499-6

Published by Kluwer Academic Publishers,
P.O. Box 17, 3300 AA Dordrecht, The Netherlands.

Kluwer Academic Publishers incorporates
the publishing programmes of
D. Reidel, Martinus Nijhoff, Dr W. Junk and MTP Press.

Sold and distributed in the U.S.A. and Canada
by Kluwer Academic Publishers,
101 Philip Drive, Norwell, MA 02061, U.S.A.

In all other countries, sold and distributed
by Kluwer Academic Publishers Group,
P.O. Box 322, 3300 AH Dordrecht, The Netherlands.

Printed on acid-free paper

TABLE OF CONTENTS

Part III : Contributed papers

WELCOMING ADDRESS

First of all, I want to welcome all participants to this Workshop, in particular those who accepted to give a lecture. Such qualified and diversified participation underlines the interest and importance of the theme.

This is the second of Workshops held here at Istituto Guido Donegani. It belongs to the mission of IGD to explore frontier technologies and themes, both by fundamental and application point of view, in order to help the Company to maintain and improve its position and to open new research fields susceptible of industrial application.

I hope you all will have a fruitful time participating to the Workshop and subsequent free discussions. Thank you again for your participation and best wishes for your scientific activity.

STEFANO CAMPOLMI
(General Manager)

INTRODUCTORY REMARKS

The use of large amounts of energy has always fascinated humans. In ancient cultures the thunderbolt (typical example of high energy density phenomenon) was considered an instrument of divinity. For this reason we chose the silhouette of Zeus hurling a thunderbolt (from a Vth century BC Greek statuette) as the logo of this book.

Back to our times, we meant "High Energy Density Technologies in Materials Science" as a number of techniques which allow to transform materials in non-equilibrium conditions, obtained using strong excesses of energy. The energy transfer is performed by directional beams (electrons, photons, ions) or by electric discharges, creating plasma regions.

There have been two main reasons for the organization of this Workshop at Istituto Guido Donegani, a corporate research center mainly involved in chemistry research. The first is to continue the series of IGD Scientific Workshops started in 1987 with "Strategies in Computer Chemistry". The second and most important reason is that we think that several technologies invented for "high-tech" sectors of industry could be advantageously used in other areas. In other words, we felt cross-fertilization between high energy density technologies and chemistry could prove very fruitful.

We tried to put together a Workshop program able to suggest technology transfer and to open minds to new ideas and applications. Our keynote lectures are particularly indicative in this respect. J. Bargon, from the University of Bonn (FRG), introduced the topic of high energy density technologies in microelectronics, with particular emphasis to lithography, an area which has seen one the most important efforts in cross-fertilization between different areas of knowledge so far. L. Wiesner, BGS (FRG), contributed on a high energy density application which has already gained a firm status in the industrial environment, using electron beams for crosslinking of polymers. H. Yasuda, from the University of Missouri (USA), provided a very valuable insight in the application of cold plasma techniques to polymers, which is becoming very important in industrially important areas (packaging, automotive, biomaterials, etc.).

We had then tutorial lectures providing fresh ideas in some important applications of high energy density technologies. M. Boulos (University of Sherbrooke, Canada) gave a wide survey on the application of thermal plasmas to the production of ceramic coatings. R. d'Agostino (University of Bari, Italy) discussed the kinetics of plasma deposition and etching of fluoropolymers. L. Martinu (Charles University, Czechoslovakia) reviewed the recently acquired knowledge on deposition and applications of amorphous carbon films. His lecture was complemented by that of E. Kay (IBM Almaden Research Center, USA), not published in these Proceedings, on the synthesis and physical properties of plasma-polymers containing metal clusters.

Our beam session included first of all a talk about the interaction of electron beams with materials, given by A. Luches (University of Lecce, Italy). G. Foti (University of Catania, Italy) provided a fresh insight into physics and applications of ion implantation in polymers (not published in these Proceedings, substituted by a paper by ourselves). E. Occhiello (Istituto Guido Donegani, Italy) reviewed the applications of UV lasers to the treatment of polymers. W. Cerri (CISE, Italy) concluded examining the use of CO_2 lasers in the treatment of metals and ceramic materials (not included in these proceedings).

Besides invited lectures, six more contributions have been presented, most of them from industry research people, confirming the interest created by the Workshop.

We hope that lectures and discussions during the Workshop helped people of different origin to achieve new ideas and instruments for their own job. Certainly it was so for us and our colleagues at IGD.

We thank the Direction, colleagues and collaborators in IGD for their support and help, without their essential contribution the workshop would not have been possible. We want finally to acknowledge the patient and accurate revision of manuscripts and illustrations by Mrs. D. Defilippi Occhiello, who is also to be credited for the cover drawing.

F. GARBASSI
E. OCCHIELLO

Part I

KEYNOTE LECTURES

LITHOGRAPHY FOR HIGH ENERGY TECHNOLOGY IN MICROELECTRONICS

Joachim Bargon

Institut für Physikalische Chemie
der Universität Bonn
Wegelerstr. 12, 5300 Bonn 1, West-Germany

Abstract

The success of the ever increasing integration in high energy density technology strongly depends upon the lithographic techniques, in particular on the reproducibility of the smallest line dimension and the associated yield. Different lithographic concepts are compared and photolithography is shown to dominate the field. UV-lithography allows the fabrication of linewidths well below 0.5 µm. X-lithography has potential, but its success depends upon the availability of compact synchrotrons, while e-beam lithography is essential for mask-making.

Introduction

The remarkable increase in integration density of microelectronic density is due to a number of reasons. Among them are simplification of the circuitry and a substantial reduction of the minimum feature size, which is required to store information, for example in a dynamic random access memory (DRAM) chip. During the past two decades the minimum feature size has decreased from about 8 µm in the 1 kbit RAM chip in 1975 to 0.85 µm in the 4M DRAM introduced in 1988 (Figures 1-4).

The increase in integration density and the associated increase in lithographic resolution has been achieved through the use of both higher aperture lenses and shorter wavelength. Thus, whereas the optical apertures for lenses using the 436 nm line of high pressure mercury arc lamps (the so-called g-line) was around 0.3 in 1982, modern systems achieved 0.54 in 1988 while decreasing aberations, distortions and field curvature (2).

As is well known from microscopy, the resolution limit of an imaging system is a function of the wavelength of the exposing radiation: the shorter the wavelength, the higher the resolution which can be achieved. Consequently, the push to a higher degree of integration drives the exposure wavelength of lithographic systems to ever shorter wavelengths. The desire to increase resolution while maintaining the depth of focus of the exposure system has led to the shorter wavelength of the so-called i-line of the mercury arc spectrum located at 365 nm. The use of 200 to 300 nm UV-light to extend the resolution of photolithography was

3

F. Garbassi and E. Occhiello (eds.), High Energy Density Technologies in Materials Science, 3-17.
© 1990 *Kluwer Academic Publishers.*

4

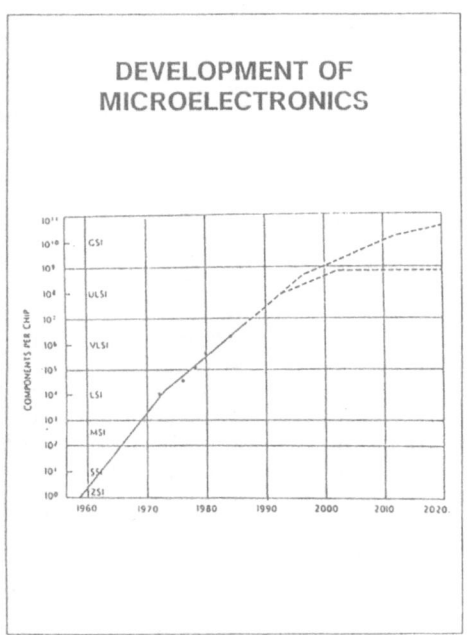

Figure 1

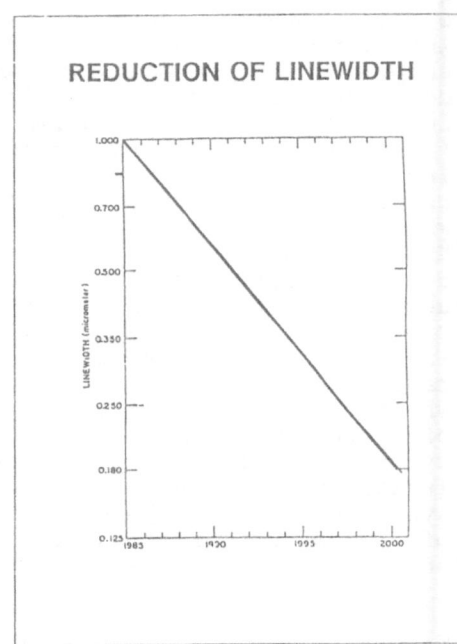

Figure 2

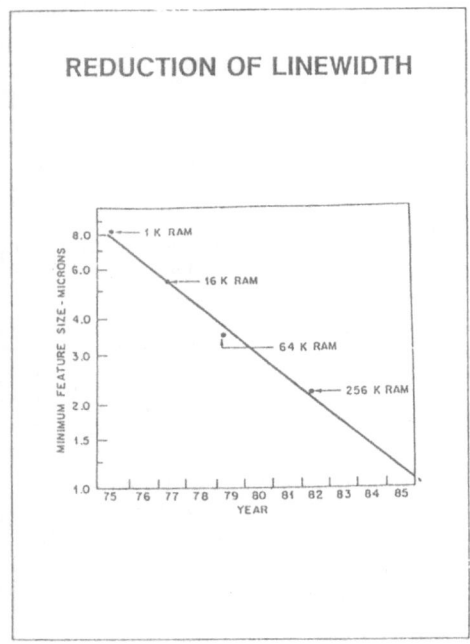

Figure 3

Figure 4

first reported by Moreau and Schmidt in 1970 (3), who used poly(methyl-methacrylate) as a resist and described most of the features of deep UV-lithography (DUV).

This technique was further redefined by Lin (4), who clearly demonstrated the submicron resolution capability of DUV.

The advent of excimer lasers has made it attractive to take advantage of their intense (and monochromatic) radiation not only in the DUV-region, but to extend it below the 200 nm regime. Feldman and co-workers have reported results obtained using poly(butene - 1 - sulfone) exposed at 185 nm (6).

Parallel to the development of more sophisticated optical systems and more powerful light sources, the photosensitive layers have been modified to match the push to shorter wavelength and thus to higher resolution. These photosensitive organic coatings are called "resists" or "photoresists".

Conventional Resist Technology

Resists are used in microelectronic processing technology to transfer photoexposed patterns into suitable substrates, mostly silicon wafers. They are typically organic polymers, which are applied as coatings form solution and resist etching.

Depending on their response to the exposing radiation they can either be classified as positive or negative resists, expressions which stem from the printing technology, where this concept of photo-lithography has been used long before the advent of microelectronics (1).

Upon exposure to radiation positive resists become more soluble in the irradiated regions relative to the unexposed ones, whereas the reverse is true for negative resists, i.e. they become less soluble (Figures 5 - 8).

Positive Resists

Conventional positive photoresists consist of a matrix resin and a photoactive component (PAC). Typically the matrix resin is a condensation product derived from phenol or m-cresol and for-maldehyde, whereas the PAC is a substituted diazonaphthoquinone (7).

The PAC acts as a dissolution inhibitor, which upon exposure decomposes into a base-soluble form and then functions as a dis-solution promoter. As such the PAC is a photo-addressed switch. In the irradiated areas the PAC cleaves off nitrogen and under-goes a rearrangement, known as the Wolff rearrangement to ini-tially yield a ketene, which reacts with water and forms a base-soluble indenecarboxylic acid (8). This way the irradiated areas are rendered more soluble, resulting in a positive tone image. Photoresists of this positive kind have accounted for the vast majority of resist applications, because they have been adapted to a significant number of processing changes, for example from wet etching using hot hydrofluoric acid to dry etching in a CF_4 plasma.

Negative Resists

Negative resists on the other hand are based on an entire different principle of operation, namely on the generation of cross-linkages, which render the exposed areas insoluble.

One of the first synthetic negative photoresists used for microelectronic circuitry was the Kodak Photosensitive Resist (KRP), which is available as a solution of poly(vinyl-cinnamate).

Upon the exposure the cinnamate side groups dimerize and form truxillic or trucinic acid-type cross-linkages. KPR is typically further sensitized to the appropriate wavelength, for example by a diaza-benzathrone.

Another widespread type of negative photoresist is based upon cyclized poly(isoprene),which is rendered photosensitive via admixed bis-aryldiazides (10). Typical examples are the Kodak Metal Etch Resist (KMER) or the closely related Kodak Thin Film Resist (KTFR).

Negative resists suffer from swelling, which results in a lower resolution than that of their positive competitors. Furthermore, they typically experience a loss of film thickness during the development process.

Another concern, namely to save on processing steps in an effort to increase the yield of usable circuits has led to the concept of the "permanent resist", whereby the developed resist is left behind as an organic insulator (11). These systems typically consist of modified, i.e. functionalized polyimides. A characteristic example is the Selectilux family of resists (2).

High Density Lithographic Techniques

Among the types of lithographic techniques which provide higher resolution than the conventional photolithography operating in the visible part of the spectrum, UV-lithography has succeeded in displacing the other competitors such as electron beam-, ion beam-, and X-ray lithography because of its relative ease and convenience. Currently, a considerable fraction of the lithography used in production lines operates in the so-called "near UV" domain, i.e. at wavelengths between 350 and 450 nm to about 250 nm, even though the range 200 to 300 is also called "deep UV-lithography" (4).

Concurrent with the development of more potent exposure tools for optical lithography, resist system operating in the mid- and deep UV range had to be developed.

The conventional, commercially available resists such as AZ1350J and equivalent formulations show poor performance in the mid- and deep UV compared to the near UV, chiefly because of their drastically lower sensitivity. This decrease in sensitivity has mostly 3 reasons:

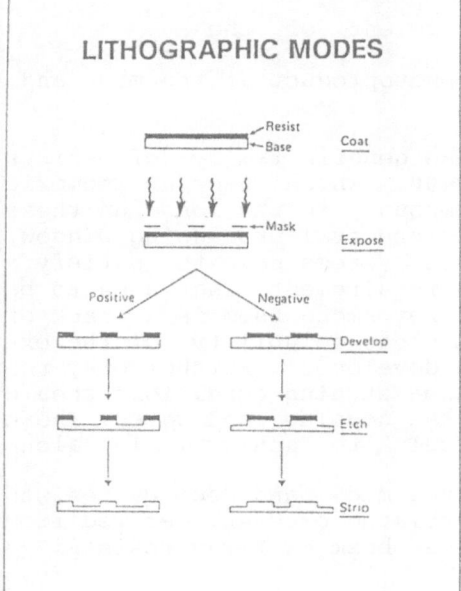

LITHOGRAPHIC MODES

Resist
Base — Coat

Mask
Expose

Positive — Negative

Develop

Etch

Strip

Figure 5

CONVENTIONAL RESIST SYSTEMS

- NEGATIVE RESISTS:
  → Matrix Resin:
    * Cyclized Rubber
    * Novolac
    * Poly(p-hydroxystyrene)
  → Photoactive Component
    * Bisazide
- POSITIVE RESISTS:
  → Matrix Resin:
    * Novolac
    * Poly(p-hydroxystyrene)
  → Photoactive Component
    * Diazoquinone

Figure 6

NEGATIVE RESIST

- MATRIX RESIN:
  → Cyclized Rubber

$$CH_2=C(CH_3)-CH=CH_2 \xrightarrow{Z.N.} -[CH_2-C(CH_3)=CH-CH_2]-$$

$$\xrightarrow{H^+} \text{Cyclized Rubber}$$

Cyclized Rubber

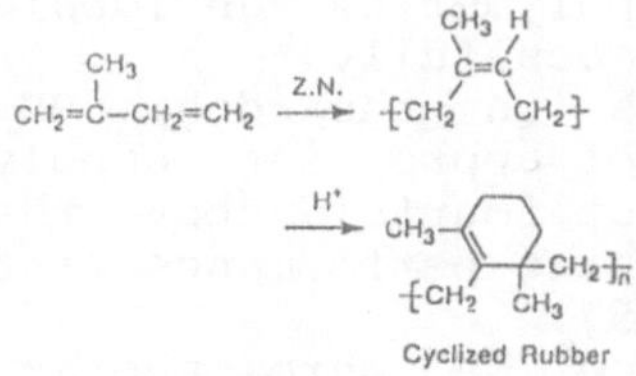

- PHOTOACTIVE COMPOUND
  → Bis-arylazide

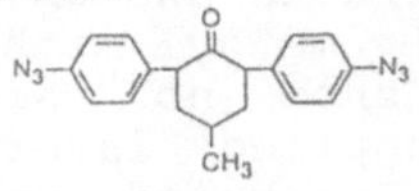

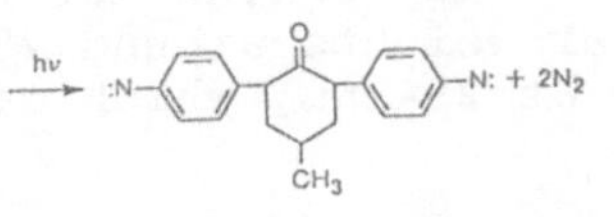

$$N_3-\text{...}-N_3$$

Bis-arylazide

$$\xrightarrow{h\nu} :N-\text{...}-N: + 2N_2$$

Figure 7

POSITIVE RESIST ACTION

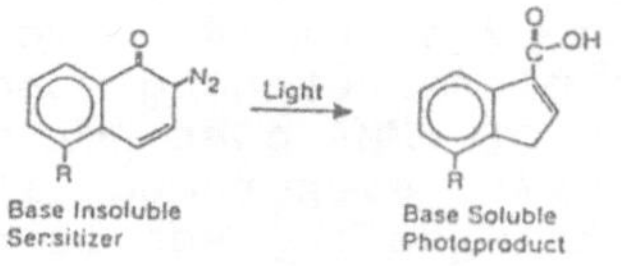

$$\xrightarrow{Light}$$

Base Insoluble Sensitizer — Base Soluble Photoproduct

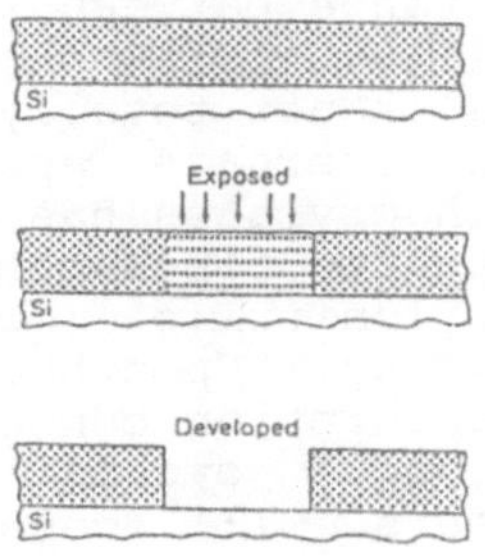

Si

Exposed

Si

Developed

Si

Figure 8

1. The self-absorption of UV irradiation by the novolac resin system.
2. The lower molar extinction coefficient of the photoactive compound at 313 nm.
3. The residual absorption of the photoproduct at the mid- and deep UV wavelengths.

Though improved versions of these generic family of resists are available, for example the AZ2400, which uses an isomeric form of the AZ1350J's photoactive compound, in the long run these systems have reached a limit of their spectral processing window.

Alternate approaches to resist systems have to satisfy a relatively long list of simultaneous requirements that used to be matched by the novolac-type family. Key among them is a lack of swelling in the unexposed parts but clean solubility of the exposed areas in aqueous base as the developer. Furthermore, the resistance of the new resists to plasma etching conditions should resemble - if not surpass - that of the novolac candidates. These requirements alone are already difficult to achieve, let alone additional processing requirements.

A variety of concepts for novel mid- and deep UV resists have been taken, some of which are related or even derived from approaches taken to obtain sensitive e- beam or X-ray resists.

Positive Deep UV Resists

Several positive working resist families have been identified and investigated, of which only a few characteristic examples will be outlined in the following.

1. Methacrylate Resists

Poly(methyl methacrylate), PMMA, homopolymers absorb UV light with a maximum absorption coefficient of about 0.45 μm^{-1} at around 215 nm (4). Together with the poor performance of PMMA under plasma etching conditions (13) the relatively low UV sensitivity of PMMA does not qualify the homopolymer as an ideal deep UV resist, even though it may be used successfully.

Efforts to sensitize methacrylates in the deep UV have succeeded in increasing the sensitivity of appropriate copolymers in the range between 230 and 260 nm about fourfold above that of PMMA (14). Unfortunately, the plasma etching performance of these systems is not totally satisfactorily (15).

An alternate approach to introduce UV chromophores into polymethacrylates has been chosen by Hiraoka and co-workers (15). They applied copolymers consisting of methyl methacrylate and methacrylonitrile, which initially show no significant UV sensitivity. Upon heating of already applied layers, however, a broad absorption centered at about 245 nm builds up, due to the formation of cyclic structures in the side chains of these copolymers. 1:1 copolymers with molecular weights around 250000 display a sensitivity of about 50 mJ/cm^2 at 254 nm, which can be sensitized to about 10-20 mJ/cm2 using p-t-butyl benzoic acid as the sensitizer (15). These systems have an acceptable resistance to plasma etching conditions in a CF_4/O_2 -plasma which can be further improved upon postbaking exposed samples prior to the development (15).

A system related to PMMA is poly(methyl isopropenyl ketone), PMIPK (16). It absorbs UV light shorter than 340 nm and is about 5 more sensitive than PMMA. Its plasma etching resistance is worse than that of PMMA, however (15). Nevertheless, PMIPK can be sensitized by an additional factor of 3, and it is available under the trade name ODUR from Tokyo Ohka Kogyo Co of Japan (17). Similarly poly (isopropenyl tert-butyl ketone) PIPTBK is 12 times more sensitive than PMIPK (18).

Another type of UV resist containing the methacrylate polymer backbone is poly(methyl glutarimide), PMGI. Originally developed as a high temperature e-beam resist (19, 20), PMGI was subsequently shown (21) to be a useful UV resist, comparable to PMMA. This polymer is sensitive below 280 nm, resistant to common organic solvents but soluble in aqueous base and thermally stable to about 185 °C. The material is offered commercially by Shipley Co (22, 23) and used in multilayers resist formulations as a thick planarizing layer, as will be discussed later.

2. Poly(olefin sulfones)

Originally developed as e-beam resists (24), poly(olefin sulfones), obtained as copolymers of sulfur dioxide and appropriate olefins, represent another class of candidates that absorb UV light in the domain 200 to 300 nm.

3. Photocatalytic Systems

New deep UV resist materials based upon the acid-introduced cleavage of polymers have been developed by Ito and co-workers (25,30). These systems provide greatly increased sensitivity without significant loss of resolution. The principle underlying this novel class of resists is that of chemical amplification (31). For this purpose an onium salt is photochemically decomposed to yield an acid which subsequently induces a variety of catalytic reactions. Originally developed for the photoinduced curing of epoxy resins (32), these systems have been extended to both self-developing (27,28) and thermally developable resist materials (33), some of which work as so called dual-tone materials (27,30). Mixed with novolac resins, the above photoacid generators have been found to act as photoactive dissolution inhibitors (25), much like the well known diazonaphthoquinones. Perhaps due to the high efficiency of the photoacid generators, formulations based upon novolacs can be used as rather sensitive deep UV resists despite the high absorption of novolac resins at 254 nm (34), (Figures 9 - 12). It appears that systems based upon chemical amplification can increase the sensitivity of resists by orders of magnitude without a significant sacrifice of resolution (35).

ACIDOLYSIS OF POLYMERS

- Acid Induced Hydrolysis of Chemical Bonds

$$S \xrightarrow{h\nu} A$$

$$\underset{\substack{| \\ \text{OCOBu}^t \\ \| \\ O}}{-CH_2\overset{R}{\underset{|}{C}}-} \xrightarrow{A} \underset{\substack{| \\ \text{OH}}}{-CH_2\overset{R}{\underset{|}{C}}-}$$

Figure 9

DRY DEVELOPMENT

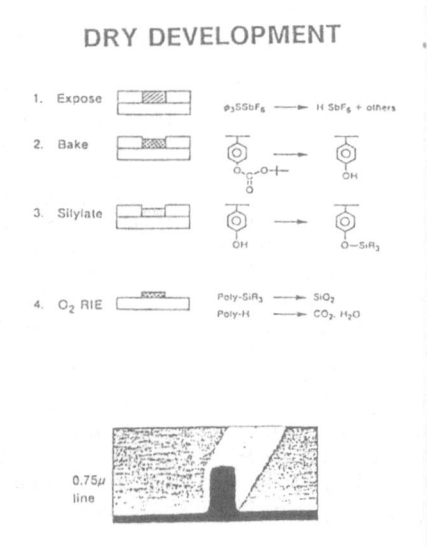

Figure 10

ACIDOLYSIS OF POLYMERS

- Acid Induced Hydrolysis of Chemical Bonds

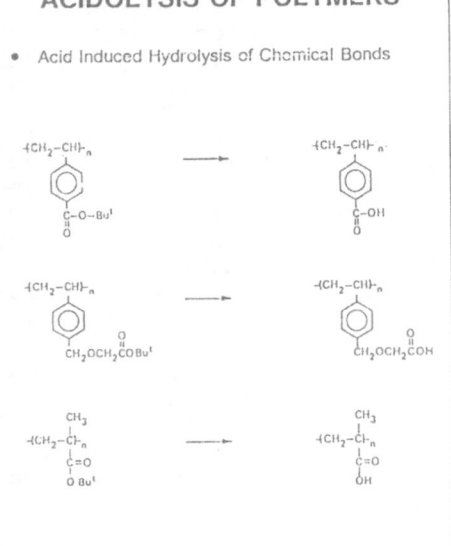

Figure 11

ACIDOLYSIS OF POLYMERS

- Acid Induced Hydrolysis of Chemical Bonds

$$ArN_2^+ \, MX_n^- \xrightarrow{h\nu} ArX + N_2 + \underline{MX_{n-1}}$$

$$Ar_2I^+ \, MX_n^- \xrightarrow{h\nu} ArI + \underline{HMX_n} + others$$

$$Ar_3S^+ \, MX_n^- \xrightarrow{h\nu} Ar_2S + \underline{HMX_n} + others$$

$$ArCOCH_2SR_2 \, MX_n^- \xrightarrow{h\nu} ArCOCH=SR_2 + HMX_n$$

$$MX_n = BF_4^-, \, PF_6^-, \, As_6^-, \, SbF_6^- \text{ etc.}$$

Figure 12

Contrast Enhancement

Efforts to push the resolution limit of photolithography include to compensate the physical consequences of diffraction via a non-linear resist response. One approach to improve the diazonaphthoquinone (DAQ) novolac system has taken advantage of the fact that a multifunctional photoactive compound containing more than one DAQ unit gives rise to a non-linear contrast response curve (Figure 14). The chemistry, contrast and resulting line profiles are shown in Figures 13 - 15. The line profile improves significantly if multifunctional photoactive compounds are used (Figure 15) (36).

An alternate approach to achieve a higher contrast, i.e a higher resolution at a given wavelength, is based upon non-linearity due to a bleachable dye (37). These dyes originally absorb very strongly at the exposure wavelength, but yield a photoproduct, which no longer does. These dyes are applied via spin coating over the photoresist prior to its exposure.

Since the bleaching is a function of irradiation intensity, the center of the exposed area bleaches more effectively reduces the minimum exposure dose to achieve a usable contrast. Since the exposure of the photoresist occurs only after bleaching the 100 to 300 nm thick dye top layer, the exposure times have to be increased about two- to three-fold. Therefore, this process increases resolution but reduces throughput (38).

The concept of using a photobleachable dye in form of a contrast enhancement layer significantly improves the resolution potential of photolithography as outlined in Figures 16 and 17.

The chemistry of different classes of dyes which have successfully been used for contrast enhancement layers is outlined in Figure 18. They include nitrones (37) as originally proposed by General Electric, diazonium salts (39, 40) as suggested by Hitachi and Matsuchita, and polysilanes used by IBM (41).

Image Reversal

An additional improvement of convential novolac/ diazonaphthoquinone resists can be achieved by adding an appropriate base to the resist formulation. Such systems have been described by McDonald et al. (42) and by Moritz (43). A typical base for this purpose may be an alkylamine or melamine. The process is outlined in Figures 19 and 20. It allows image reversal depending on the sequence of processing steps as outlined in Figure 19.

Other Types of Lithography

The progress achieved using conventional photolithography has limited the advances of alternate lithographic techniques, such as X-ray, e-beam and ion beam lithography (Figure 21). All of these competitors are basically more cumbersome to apply, with

12

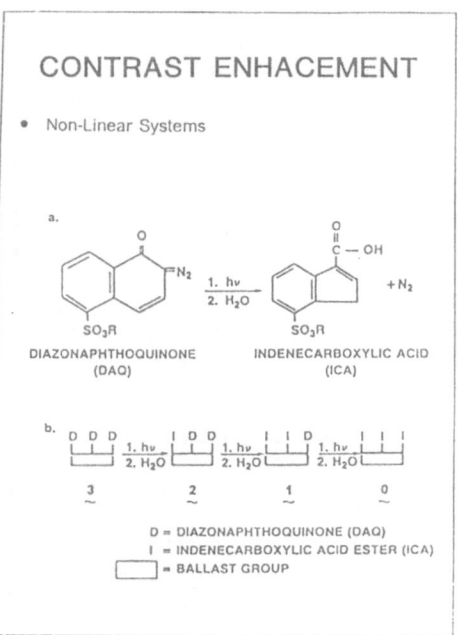

Figure 13

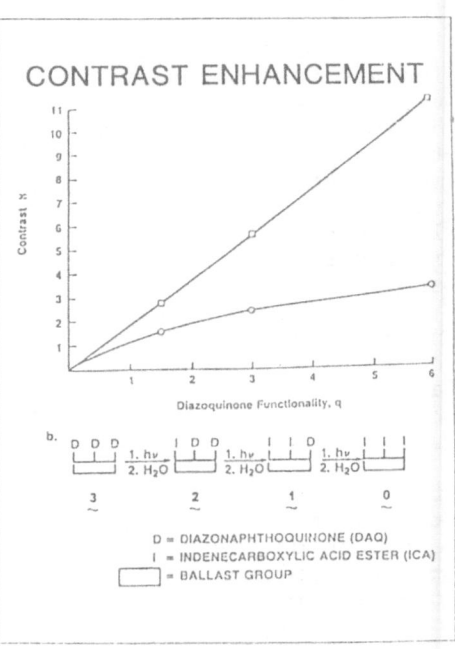

Figure 14

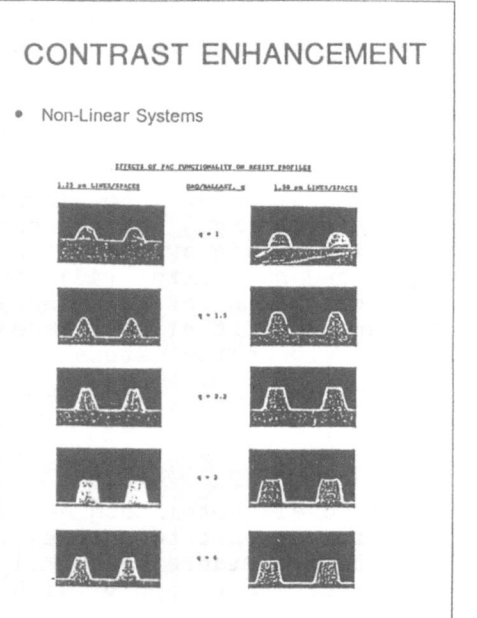

Figure 15

CONTRAST ENHACEMENT

- Contrast Enhancement Layer (CEL)

  A photobleachable dye imbedded in a layer on top of the photoresist renders the system more non-linear, bleaching in the heavily exposed areas, but retaining some absorption in the less exposed zones.

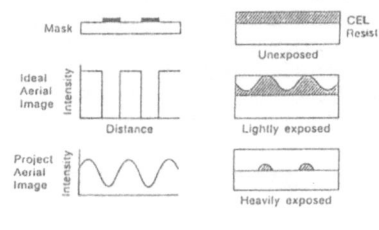

  → The residual absorption leads to increased contrast

Figure 16

13

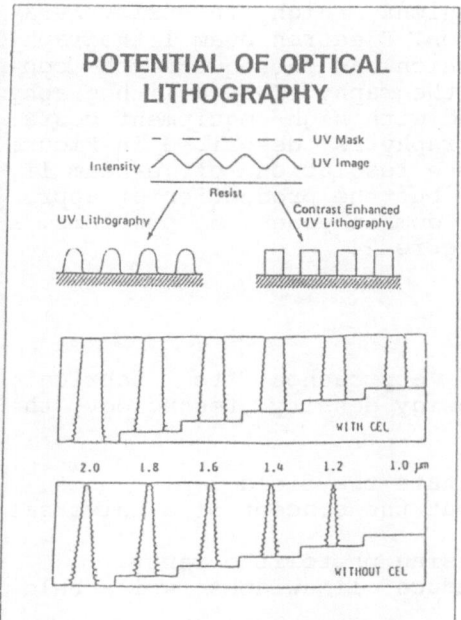

Figure 17

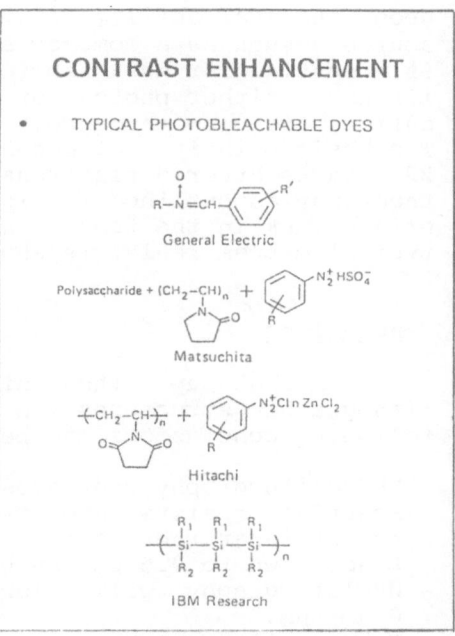

Figure 18

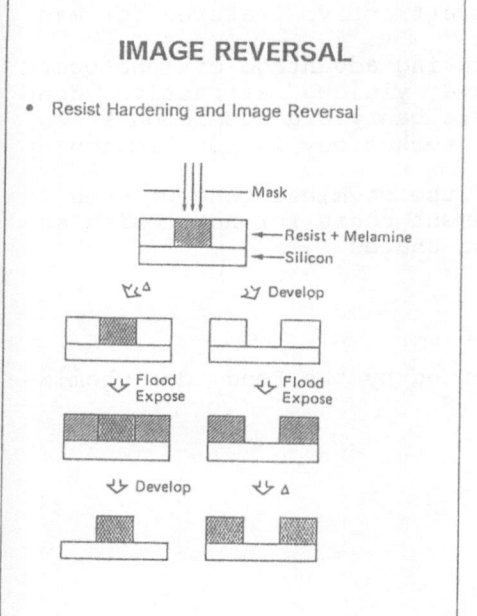

Figure 19

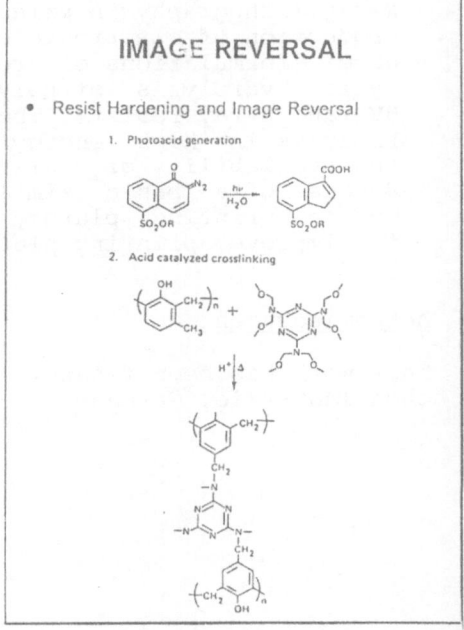

Figure 20

the possible exception of X-rey lithography. The latter depends upon the availability of a convenient, high intensity X-ray source, such as a compact synchrotron. Electron beam lithography is essential for mask fabrication, which can subsequently be copied using either photo- or X-ray lithography. E-beam lithography suffers from low throughput combined with high equipment costs. The basic principle of e-beam lithography is described in Figure 22. Backscattered electrons limit the resolution of e-beam lithography as outlined in Figure 23, but the production of appropriate mask in the tens of a micron domain poses no problems. A typical e-beam resist is shown in Figure 24.

Conclusions

Summing up the different approaches to fabricate lithographic structures in high energy density technology, the following conclusions can be drawn:

- Photolithography continues to dominate the field.
- Positive resists are favored, but the concept of a universal resist is attractive.
- Lines down to 0.5 μm can be made using photolithography.
- UV-lithography will allow to reduce linewidths well below 0.5 μm.
- E-beam lithography will be used for mask making, and in isolated cases direct-write e-beam techniques may be used for customized chips with limited editions.
- X-ray lithography has potential but hinges on theavailability of convenient high intensity X-ray sources, such as compact synchrotrons.
- X-ray lithography in principle has attractive features for mass production of electronic circuitry.
- Novel formulations of resists taking advantage of photocatalytic hydrolysis of polymers have yielded attractive deep UV- and X-ray resists. These resists can yield submicron lines.
- Progress in high energy density technology is not limited by the availability of resists.
- Because of a basic similarity of the chemical concept used in modern printing plates, development costs for new resist and for improved printing plates can be shared.

Acknowledgement

This work has been financially supported by the Fond der Chemischen Industrie, Germany.

References

1. H.W. Vollmann, Angew. Chem., 92, 95-99 (1980);
    Angew. Chem. Int. Ed. Engl., 19, 99-110 (1980) and references
    therein.

2a. J.H. MaCoy, W. Lee and G.L. Varnell, Solid State Tecnol.3,
    87-91 (1989) and references therein.
2b. W. Lee, J.T. Lyon, J.H. MaCoy, G.L. Varnell. Proc. SPIE
    Vol.921, 312 (1988).

3. W.M. Moreau and R.R. Schmidt, 138th Electrochem. Soc.
    Meeting, Extended Abstracts, p. 459 (1970).

4. B.J. Lin, J. Vac. Sci. Technol., 12, 1317-1320 (1975)

5. K. Jain, Proc. Int. Conf. Microcircuit Eng., p. 49 (1983)

6. M. Feldman, D.L. White, E.A. Chandross, M.J. Bowden and J.
    Appelbaum, Proc. Kodak Microelectronics Seminar 40 (1970).

7. J. Bargon in "Methods and Materials in Microelectronic Tech-
    nology", J. Bargon ed., Plenum Press, New York (1984).

8. O. Sues, Justus Liebigs An.Chem., 556, 65 (1944).

9. K.G. Clarke in "Negatives Photoresists", R.J. Cox, ed.
    AcademicPress, London-New York, pp. 249-273 (1975)

10. R.F.M. Thornley and T. Sun, J.Electrochem.Soc., 112, 1151
    (1965)

11. R. Rubner, H. Ahne, E. Kühne and G. Koldozieji,
    Photograph. Sci. Eng., 23, 303 (1979).

12. H.J. Merrem, R. King and H. Härtner, E. Merck, Darmstadt,
    West Germany.

13. L.A. Pederson, J. Electrochem. Soc., 129, 205 (1982).

14. E. Reichmanis and C.W. Wilkins,Jr., ACS Symposium Series
    184, Washington, D.C., pp. 29-43 (1982).

15. H. Hiraoka, W.L. Welsh, Jr. and J. Bargon,
    J. Vac. Sci. Technol., B1, 1062-1065 (1983).

16. A. Levine, Soc. Plastics Eng., tech. pap. (1973) p: 106.

17. M. Tsuda, S. Oikawa, Y. Nakamura, H. Nagata, A. Yokata, H.
    Nakane and T. Mifune, Photographic Science and Eng., 23,
    290-296 (1979)

18. S.A. MacDonald, H. Ito, C.G. Willson, J.W. Moore, H.M. Gharapetian and J.E. Guillet, ACS Symposium Series 266, "Materials for Microlithography", L.F. Thompson, C.G. Willson and J.M.J. Frechet, eds., ACS, Washington, D.C. (1984), p.179.

19. J Bargon, E. Gipstein and H. Hiraoka, US Patent 3,964,908 (1976).

20. J. Bargon, E. Gipstein, J. Bargon and L.W. Welsh, Jr., J. Appl. Polym. Sci., 22, 3397 (1978).

21. H. Hiraoka, Macromolecules, 10, 719 (1977).

22. M.P. de Grandpre, D.A. Vidusek and M.W. Legenza, SPIE, Vol.539, 103 (1985).

23. M.W. Legenza, D.A. Vidusek, M.P. de Grandpre, SPIE, Vol.539, 250 (1985).

24. T. Bowmer and J. O'Donnell, J. Polym. Sci.Chem.Ed. 19, 45 (1981).

25. H. Ito and E. Flores, J. Electrochem. Soc., in press, IBM Research Report 5961 (1987).

26. H. Ito, E. Flores and A.F. Renaldo, ibid., in press, IBM Research Report 5962 (1987).

27. H. Ito and C.G. Willson, Polym. Eng. Sci. 23, 1012 (1983).

28. H. Ito and C.G. Willson in "Polymers in Electronics", T. Davidson, ed., ACS Symposium Series 242, ACS, Washington, D.C. (1984), p.11.

29. H. Ito, C.G Willson and J.M.J. Frechet, SPIE 771, 24 (1987).

30. H. Ito and M. Ueda, Macromolecules in press.

31. C.G. Willson, H. Ito, J.M.J. Frechet, T.G. Tessier and F.M. Houlihan, J. Electrochem. Soc. 133, 181 (1986).

32. J.V. Crivello in "UV-Curing: Science and Technology", S.P. Pappas, Ed., Technology Marketing Corporation, Stanford (1978), pp.23-77.

33. H. Ito and R. Schwalm, J. Electrochem. Soc., in press.

34. D.R. McKean, S.A. MacDonald, N.J. Clecak and C.G. Willson, SPIE (1988) in press.

35. T.Twayanagi, T.Kohashi, S. Nonogaki, T Matsuzawa, K. Donta and H.Yanazawa, IEEE Trans. Electron. Devices, ED 28, 1306 (1981).

36. R. Trefonas III and B.K. Daniels, Proc. SPIE 771, 194-210 (1987).

37a.B.F. Griffing and P.R. West, Semiconductor Int. (1983), pp.17-18.
37b P.R. West, C.G. Davis and B.F. Griffing, J. Imag. Sci., 30, 65   (1986).

38. A.R. Neureuther, D.C. Hofer and C.G. Willson, "Design of Contrast Enhancement Process for Optical Lithography", Microcircuit Engineering 84, A. Heuberger and Beneking, Eds., Academic Press, London (1985), p.53.

39. M. Sasago, M. Endo, Y. Hirai, K. Ogawa and T. Ishihare,, Proc. SPIE 631, 321 (1986).

40. S. Uchino, T. Iwayanagi, T. Ueno, M. Hashimoto and S.   Nonogaki, Proc. SPIE 771, 11 (1987).

41. D.C. Hofer, R.D. Müller, C.G, Willson and A.R. Neureuther, Proc. SPIE 469, V108 (1984).

CROSSLINKING OF INDUSTRIAL PRODUCTS BY HIGH ENERGY ELECTRON BEAMS

Lothar Wiesner

BGS Beta-Gamma-Service Dr. Wiesner GmbH & Co., D-5276 Wiehl
BGS Beta-Gamma-Service Dr. Wiesner (Süd) GmbH, D-7520 Bruchsal

Abstract

After a short review of the property changes which are produced in polymers by crosslinking, the advantages of radiation
crosslinking in comparison to chemical crosslinking techniques
are discussed. The applicability of radiation crosslinking to a
large variety of polymers and of products (cables and wires,
tubes and hoses, foam, moulded parts of different shape and size)
is stressed. Finally, economical aspects of radiation processing
and its contribution to technical innovation in the material science are outlined.

Introduction: conditions for the industrial utilization of high energy electron beams

During the past 10-15 years, an irradiation industry, based
on powerful electron accelerators and gamma sources, has
developed in some countries, particularly in the Unites States,
Japan, Germany and England for the upgrading of the properties of
a large variety of polymer products: e.g. cable and wire insula-
tions, heat shrinkable products, installation tubing, vehicle
tires, plastic films, sheet and foams and mouldel parts off different shapes and sizes. The rapid growth of the quantity of
radiation treated products which started during the seventies,
was made possible primarily by the development of reliable, economic irradiation facilities and by the increasing availability
of materials which benefit from irradiation. But at the same
time, aspects have gained in importance by the general technical
development which has favoured and continues favouring radiation
processing.
The innovation of radiation processing for manufacturing
industry is due to the unique form of introducing energy into a
material for the induction of chemical reactions. It results from
the high penetration of ionizing radiation which produces reactive species, especially radicals at ambient conditions of temperature and pressure throughout the volume of a body, i.e. oth-
erwise necessary processing conditions are avoided. This
advantage results in processing possibilities which no other
technique can offer. For example, it is possible to modify a
product only partially of differently in different · parts by in-
hibiting the radiation exposure in those areas which should not
be modified, or by a suitable choice of radiation energy which
determines the penetrated material thickness.

19

F. Garbassi and E. Occhiello (eds.), High Energy Density Technologies in Materials Science, 19–32.
© 1990 *Kluwer Academic Publishers.*

But ionizing radiation is a rather expensive form of energy. Therefore, its use is an economically feasible option only in certain areas. Indeed, all technically and economically advantageous applications of radiation belong to at least one of the following tree categories:

- Modification of polymers by crosslinking or degradation, whereby each radical produced changes on average approximately one chemical bond. As 1 kWh radiation produces about 1 mol of radicals, also 1 mol of a polymer is modified. The amount of radiation energy needed per mass unit of a product is relatively low on account of the high molecular weight of polymers.

- Processes like polymerization or graft polymerization, in which a single radical initiates a chain reaction. Thus 10^4 and even more molecules can react as a result of the generation of one radical. Again, a large mass of product results per unit of absorbed radiation energy.

- Modifications, which are so important or valuable that the cost of the radiation energy does not matter. To this category belong very different applications, such as the modification of semiconductor materials, the reduction of environmental pollution, the extension of the shelf-life of food, the sterilization of medical devices and drugs, and in principle also the radiation therapy of cancer.

Characteristics of crosslinked polymers

The crosslinking of polymer materials is - apart from eventually sterilization - at present by far the most important area of radiation processing as to the volume of materials treated, as well as for its effect on the development of new materials.

Indeed, an easily processable thermoplastic material, which is crosslinked after it has its final shape, can be considered as a new material which important properties: partly it resembles a duroplastic material, partly, at least at higher temperatures, an elastomeric material:

- Solubility and swelling in solvents are drastically reduced;

- The probability of stress cracking is decreased by several orders of magnitude;

- The crosslinked material doesn't melt anymore, thus a product retains its shape even at temperature above the melting point of the original thermoplastic polymer, as long as mechanical stresses at such temperature are relatively small. Crosslinked PE, for example, resists temperature of 250 °C for short periods of time and is suitable for long-term use at temperatures up to approximately 150 °C, if it is

appropriately   stabilized.   Similarly,   crosslinked   PVC,
polyamides and   polyesters can withstand short   exposures up
to 450 °C.

- The creeping of thermoplastic materials, which already occurs
  under mechanical   stresses at rather   low temperatures, is at
  least substantially reduced.

- The wear resistance,   the hardness as well   as other strength
  property increase.

- The   so-called   memory   effect   of   crosslinked thermoplastic
  polymers allows   the manufacture   of heat-shrinkable products
  with high shrink ratios, up to about 6:1.

The physical   process of radiation   crosslinking takes place
at low temperatures and at least primarily in the amorphous phase
of a partly crystalline polymer (1).   It originates from the formation of radicals in the polymer chain by   the abstraction   of a
hydrogen   atom,   when   ionizing   radiation   interacts   with   the
electrons in the atom shells of the irradiated material. If radicals,   produced   in   adjacent   polymer chains,   combine, at first
branching   takes   place,   then   a   three-dimensional   network   is
formed.   Thus,   the degree of   crosslinking of any   given polymer
depends on the number of radicals produced   and can   therefore be
fixed by   the radiation dose,   i.e.   the   amount   of the absorbed
radiation energy.
Polymers can also be   crosslinked by   several chemical techniques,   among which the production of radicals   by peroxides and
the   grafting   of   silanes   onto   the   polymer   and   subsequent
crosslinking in   the presence of   humidity according   to the Dow
Corning process and its variants ate the industrially most important ones.   As far as the crosslinking by peroxides is concerned,
there   are no   differences   in   the   chemistry   when   compared to
radiation crosslinking.   But products of the decomposition of the
peroxides may remain in the polymer,   and the crosslinking occurs
in the molten state when there are no crystals.   The silane process also occurs by preference in the amorphous phase   at temperatures   below the melting range of   the   crystallites.   But unlike
radiation and peroxide   crosslinking,   the polymer chains are not
coupled directly but via $Si(OH)_2-O-Si(OH)_2$-bridges. Crosslinking
knots are formed when all OH-groups of these bridges   are substituted by polymer molecules (2).
These   differences in   the distribution   and   nature   of the
crosslinks make it understandable that there are also differences
in the properties of   products crosslinked   by different methods.
An example is given in Figure 1   for polyethylene tubes   (3). The
long-term strength of the tubing   crosslinked by   electron radiation steadily increase with the degree   of   crosslinking, while a
too high degree of crosslinking reduces the long-term strength of
tubes crosslinked by peroxides or a silane process. The latter do
not achieve the performance realized   with   radiation or peroxide
crosslinking, even at the optimum gel content.

22

Long-term strength of crosslinked polyethylene tubes
(outside diameter 20 mm, wall thickness 2 mm) at 60 °C
as function of the gel content.

Fig. 1

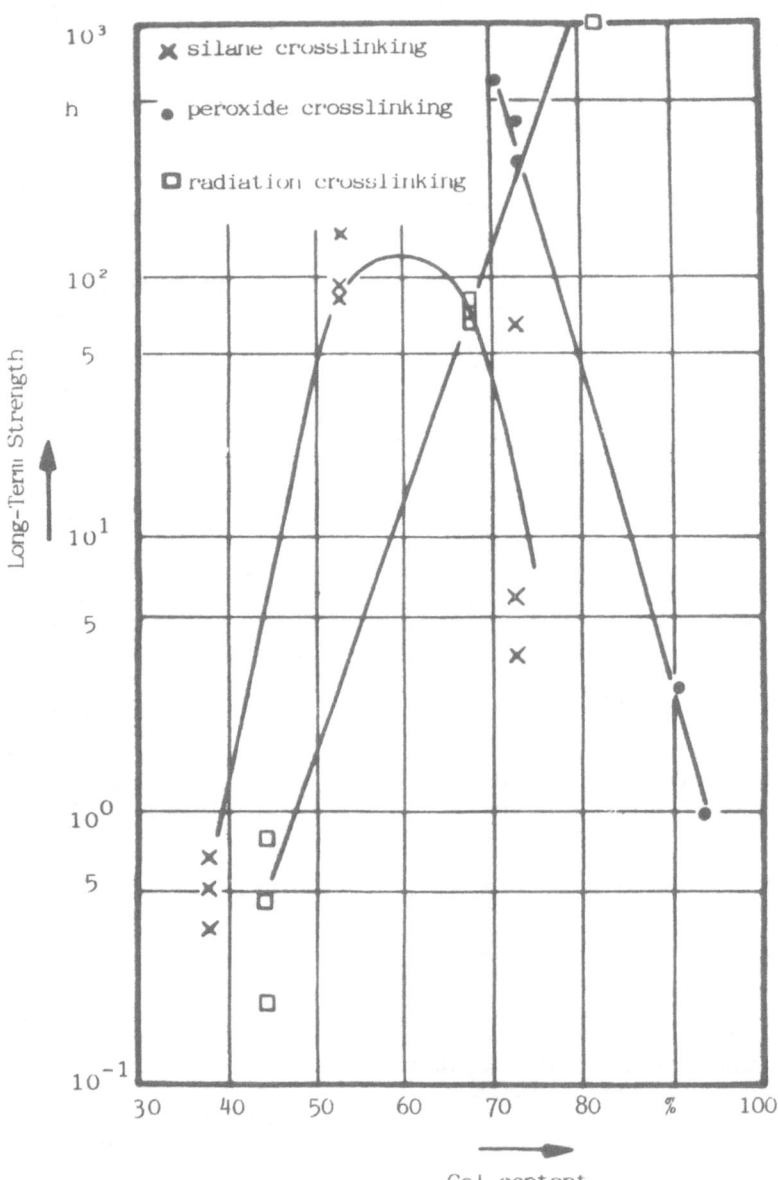

Long-Term Strength

Gel content

Fig. 2: Typical dependence of the gel content on
         the absorbed radiation energy.

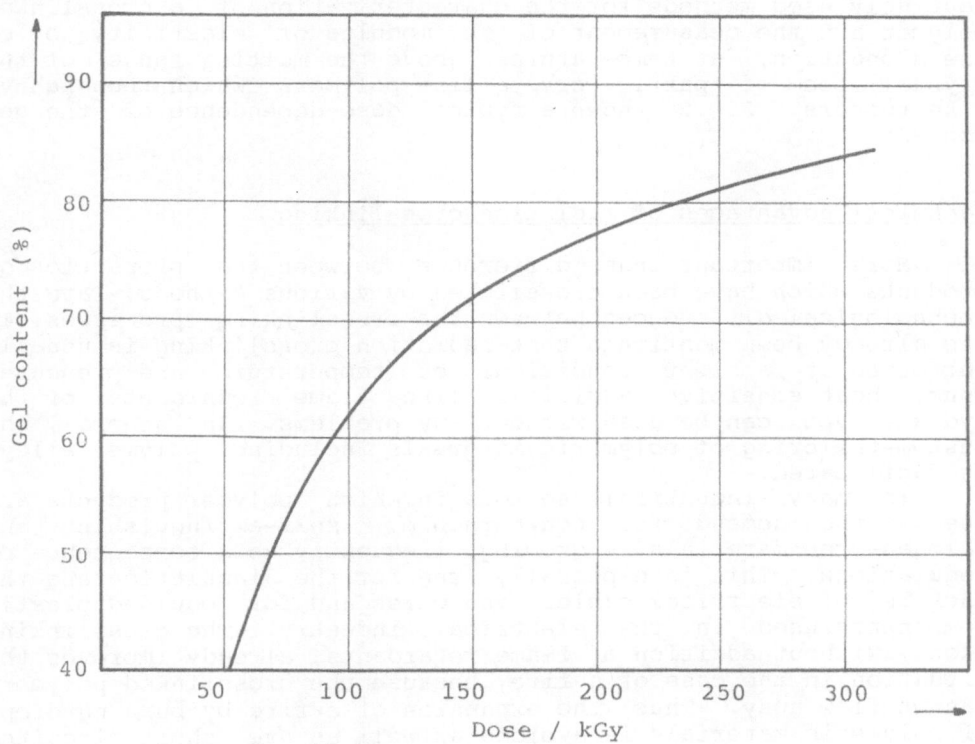

24

The degree of crosslinking is usually determined by one of the property changes induced by this process: The gel content is the fraction of a polymer which remains undissolved in a boiling solvent which readily dissolves the uncrosslinked material. Other routinely used methods for the characterization of a crosslinked polymer are the measurement of the modulus of elasticity or of the elongation, at temperatures above the melting range of the crystallites of partly crystalline polymers which then behave like rubbers. Fig.2 shown a typical dose dependence of the gel content.

Technical advantages of radiation crosslinking

More important than differences between the properties of products which have been crosslinked by various methods, are the technological differences between the crosslinking processes. It has already been mentioned that radiation crosslinking is usually performed at ambient conditions or temperature and pressure. Thus, heat sensitive additives like flame retardants of the hydrate type can be used without any problems. In general, the custom-tailoring of polymeric materials including polymer alloys is facilitated.

In many industrial sectors in which polymer products are used, the demand for non-burnable, self-extinguishing and halogen-free materials is growing, frequently as a consequence of regulations. This is especially true for the insulation and the jackets of electrical cables and wires and for moulded plastic components used in the electrical industry. The crosslinking alone, without addition of flame retardants, already improves the situation in the case of a fire, because the crosslinked polymers cannot flow away. Thus, the expansion of a fire by burning drops of polymeric materials is avoided as well as are short circuits, because the insulating material remains in its place, though eventually carbonized. Moreover, it has been found with crosslinked materials that lower concentrations of flame retardants are sufficient for the production of a non-burnable of self-extinguishing material. This is of considerable significance, because flame-retardants are not only costly additives but can also be detrimental to other properties.

Radiation crosslinking allows to easily and reproducibly adjust the degree of crosslinking to the required performance of a product by selecting the proper radiation dose. This is an important feature, because an unnecessarily high degree of crosslinking may also impair some properties: too highly crosslinked materials may become brittle and their applicability at low temperatures may be restricted due to a decreased impact strength. The possibility of fixing any desirable degree of crosslinking by the radiation dose, is of particular value for the production of high quality heat shrinkables and foams. In these applications, it is even important to reproducibility keep the degree of crosslinking within a rather narrow range in order to match the parameters of the following expansion step.

The rates of crosslinking by electron radiation are considerably higher than in the chemical processes, sometimes by orders of magnitude. BGS irradiates, for example, cables and wires, tubes and hoses at speeds up to several 100 m/min. In the crosslinking of moulded parts, a steady flow of products in boxes or on trays passes the electron beam with speeds in the order of 100 mm/s.

The separation of the shaping process of a product from the crosslinking step in the case of a radiation application, contributes to a substantial reduction of material losses, because the quality control can already be performed before the crosslinking is done. The material can be recycled if any part of the production does not meet the specification. Recycling is no more possible for crosslinked products, as the originally thermoplastic material does not melt.

Radiation processing can be organized and controlled easily as to meet the specifications regarding the crosslinking from the very beginning of a production until its termination. Thus, the formation or refuse is limited to human error, which nowadays can be eliminated for the most part by modern microprocessor controls unavoidable for technical reasons in crosslinking processes with peroxides, particularly at the start of a production.

Applicability of radiation crosslinking to different polymers

Highly significant for the irradiation process is the possibility to apply it to the crosslinking of the majority of the industrially important polymers. Table 1 lists the groups of polymers which can be crosslinked by radiation, and the most important products within each group. But their crosslinked by radiation, and the most important products within each group. But their crosslinkability varies within very wide limits, because the formation of the intermolecular bonds, being a precondition for the rise off a three-dimensional network, always competes with the degradation of the polymer by chain scission. As irradiations of industrial products are mostly carried out in the presence of air, radiation oxidation may additionally occur. The yield of each of these reactions per unit of absorbed radiation energy, called the G-value (= number of reaction/100 eV absorbed energy), depends on the molecular structure of the polymer.

These are polymers like polyethylene, copolymers of ethylene with propylene, vinyl acetate, butyl acrylate, tri- or tetrafluor-ethylene etc., polybutadiene, polysiloxane, polyvinylfluoride, which can be the crosslinking and degradation/oxidation >1 and G-value for crosslinking >1.

But even if one or both of these preconditions are not fulfilled, three-dimensional networks can still be produced in many polymers by ionizing radiation, if they are compounded with quantities of up to a few percent of polyfunctional monomers (preferably allylic or acrylic compounds), and/or by blending with a polymer, easily crosslinked by radiation (4,5,6). In the latter case, the polymers must be miscible by special procedures,

TABLE 1: Polymers which can be crosslinked by radiation
including those, needing the addition of a polyfunctional monomer for crosslinking.

Group of the Polyalkenes (Polyolefines)		G-VALUES
Polyethylene (LDPE/HDPE/LLDPE)	PE	1,4 - 1,8
Ethylene-Vinylacetate-Copolymer	EVA	2,5 - 5,0
Ethylene-Ethylacrylate-Copolymer	EEA	2,0 - 4,0
Chlorinated Polyethylene	PEC	
Chlorsulfonated Polyethylene	CSM	
Polypropylene	PP	0,2 - 0,5
Ethylene-Propylene-Diene-Elastomer	EPDM	2,6 - 3,6
Ethylene-Propylene-Elastomer	EPM	

Group of the Synthetic Rubbers		
Butadiene-Elastomer (Polybutadiene)	BR	2,0 - 5,8
Styrene-Butadiene-Elastomer	SBR	1,6 - 3,8
Acrylnitril-Butadiene-Elastomer	NBR	
Isoprene-Elastomer (Polyisoprene)	IR	0,8 - 3,5
(Polychlorbutadiene / Polychloroprene)		

Group of the Styrene Polymers		
Polystyrene	PS	0,02-0,05
Butadiene-Styrene-Copolymer	BS	
Styrene-Butadiene-Graft-Copolymer	SB	
Acrylnitril-Butadiene-Styrene		
Graft Polymer	ABS	

Group of the Halogenated Polymers		
Polyvinylchloride	PVC	0,2 - 0,3
Polyvinylidenfluoride	PVDF	
Polyvinylfluoride	PVF	

Group of the Polyvinylesters		
Polyvinylalcohol	PVAL	

Group of the Acryl Polymers		
Acrylester-Elastomer	ACM	
Polyacrylnitril	PAN	

Group of Polymers with Hetero Atoms
in the Main Chain

Polyamides	PA	

Group of the Polysiloxanes (Silocones)		
Siloxane-Elastomer (Silicone Rubber)	SI/SIR	

like grafting. Thus polyvinyl chloride, polypropylene, polyamides
polyesters, thermoplastic polyurethanes are also crosslinked by
radiation.

The addition of polyfunctional monomers may also be advantageous for the radiation crosslinking of polymers with relatively
high G-values of crosslinking for reducing the required radiation
dose and thereby the formation of degradation/oxidation products.

This is especially advisable when higher concentrations of
stabilizers and antioxidants, acting as scavengers for the radicals produced by the radiation, are added to the polymer for the
retardation of aging processes, occurring when the product is exposed to severe environmental strains for long periods of time.

The foregoing digest on the large number of possibilities
for improving the properties of the thermoplastic polymers, illustrates in rough strokes the great potential of radiation
crosslinking in the development of new and better materials for
numerous applications. Expensive materials usually also requiring
higher expenditures for bringing them into the shape of a fin-
ished product, can thereby be substituted by crosslinked thermo-
plastics, which in many cases cost only a fraction and can be
processed much more easily.

This is demonstrated by the comparison of the temperature
dependence of the surface hardness of radiation crosslinked PVC
and a conventional hard PVC (7) (Fig.3). The latter shows a steep
slope already at temperature above 90 °C. The radiation
crosslinked PVC maintains a high portion of its hardness up to at
least 176 °C, showing a behaviour expected from a truly elastomeric material in this temperature range. This performance,
achieved by radiation crosslinking of a suitable compound, broad-
ens the application spectrum of a low cost polymer like PVC.

In spite of the large quantities of products which already
benefit from a radiation treatment, up to now only a small part
of the potential offered by the radiation crosslinking of polymers is actually being exploited. This is particularly valid for
the injection moulded parts on the one hand, and for polymer alloys and fiber reinforced materials on the other hand.

The large variety of applications in the field of moulded
parts is illustrated by the spectrum of products BGS Beta-Gamma-
Services is irradiating in its two service irradiation centers in
Germany: several million pieces per month in the weight range of
up to approximately 10 g, apart from pieces with eights up to
several kg, several 1000 km of cable and wires, tubes and hoses,
and several 1000 m3 of medical devices.

Essentially the moulded parts can be described as caps, cases, gaskets, plates, plungs, sockets, spouts and stoppers in many
different forms, sometimes rather complex shapes (Fig. 4), and
for many different industrial branches, from car manufacturing
and mechanical engineering to the electrical and chemical industry and cosmetics. Mainly, they are crosslinked for reasons of
better form stability at higher temperatures and/or a higher resistance against stress cracking.

For the majority of these applications of radiation processing the development of sophisticated compounds is not required.

Fig. 4 : Examples of parts with complex shape and
wall thickness of several mm which are
routinely irradiated with electrons (7)

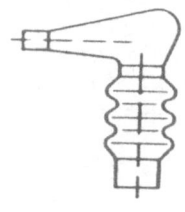
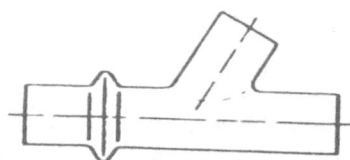

Fig. 3: Surface hardness of a conventional hard PVC compound
crosslinked by radiation as a function of temperature (7)

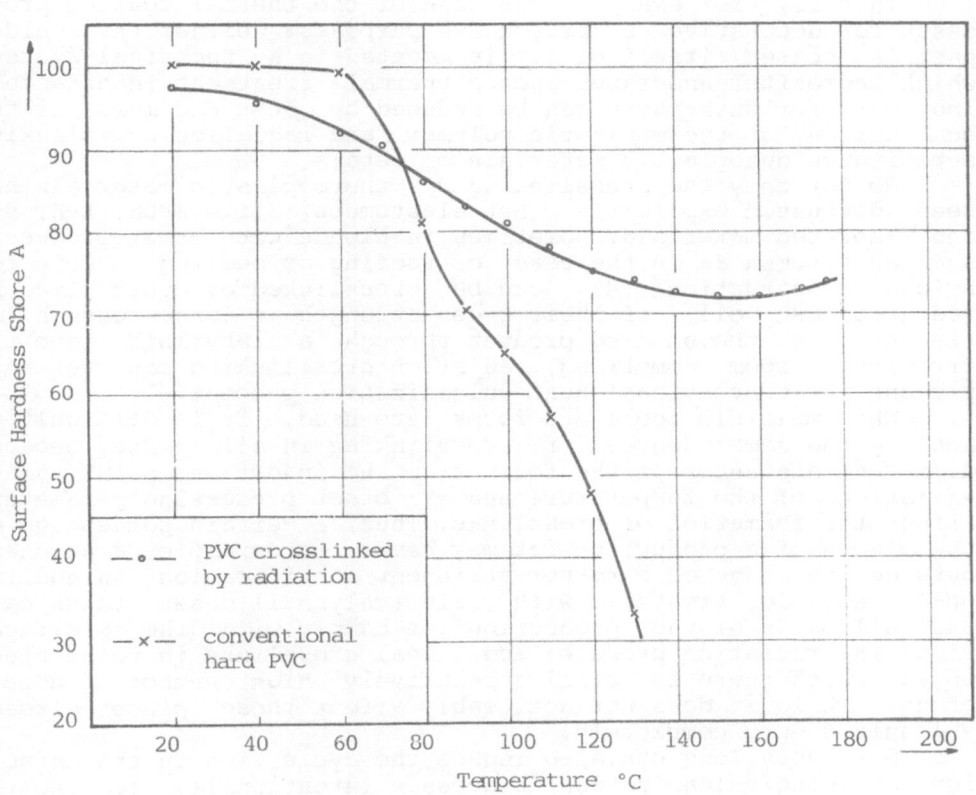

Many of the crosslinked parts have to stand temperatures above the melting point of a thermoplastic polymer only once during their lifetime, and then only for a limited time during which no noticeable aging occurs.

This is, for example, the case of the thermal coating processes for decorative or protective purposes: either the moulded part is coated itself or it is mounted in a technical system, which thereafter undergoes such a thermal treatment. Manufacturing costs for such parts can be reduced by 50% and more, if the combination of thermoplastic polymer and radiation crosslinking substitutes duroplastic materials or metals.

So far only the crosslinking of thermoplastic materials has been discussed explicitly. But elastomers, like EPDM, EPR, SBR and fluorated materials, sometimes in blends with other products, such as bitumen as in the case of roofing or sealing sheets for bridge construction (8), can be crosslinked or vulcanized by radiation as well, if their green strength is large enough for transporting the uncured product through a labyrinth into the irradiation room. Sometimes, radiation crosslinking can even supplement the conventional heat vulcanization process.

When multiple tools and forms ate used, it is difficult to achieve the same degree of crosslinking in all parts, because different distances of the forms from the injection point, small variations of the temperature and of other processing parameters affect the formation of crosslinks. Thus, a certain percentage of the pieces of a production lot may have, for example, a hardness outside the range of a rather stringent specification. An additional radiation treatment with relativelysmall doses takes care that all parts of the production lot comply with the specification: the radiation produces additional crosslinks in those pieces in which there is still a relatively high number of double bonds, while it does not noticeably affect those pieces already vulcanized more completely.

Obviously, one can also reduce the cycle time in the injection moulding/vulcanization process, intentionally leaving the crosslinking partly to the radiation treatment, which the requires correspondingly higher doses. The increase in productivity of the injection moulding machine obtain thereby, outweights the higer irradiation cost associated with the increase of the dose.

The techno-economical potential of radiation crosslinking

The technical advantages of radiation crosslinking which have been reviewed, are usually connected, at least indirectly, with economical profits. This is immediately visible if the refuse percentage of a production is reduced. As the radiation crosslinking can take place at room temperature and frequently without any additive, the total energy requirements are considerably smaller than for technologies requiring additional processing steps, expensive additives and/or the heating of the whole production to higher temperatures. A fabrication, making use of radiation crosslinking, will also in most cases occupy less space than one applying the alternative chemical crosslinking methods.

Though all these advantages have a certain economical importance, they are altogether barely decisive. So we are finally coming to the hard economical facts.

The cost per kWh of radiation energy at present is generally between 3 and 30 DM. It increases within this range with growing penetration of the radiation and decrease for facilities with larger installed irradiation capacities. But the advantages of the special form of supplying energy for a chemical reaction by radiation, outweigh in the crosslinking of polymers and in other industrial applications the high cost of the radiation energy - provided, it can be efficiently used.

The necessary expenditures for crosslinking a product are first of all determined by the required dose and by the efficiency with which the radiation energy is utilized for obtaining the specified minimum degree of crosslinking. The efficiency that can be achieved in the irradiation of a specific product depends on the suitability of a facility for that kind of product and on the homogeneity of its mass distribution in the primary beam direction.

If suitable conditions for the irradiation of a product can be chosen, and if the irradiation capacity of a modern powerful facility can be well exploit, the crosslinking cost for the majority of applications will be in the range between 2 and 5 DM/kg.

This economic feasibility is limited to facilities with large installed irradiation capacities due to two synergistic features of the radiation technology:

- The total investment required for an electron beam facility does not depend very much on the installed capacity.

- The incidence of the capital cost on the total processing cost by radiation is unusually high: it can be higher than 60%.

On account of these features of the radiation technology, service irradiation centers have been built up in many countries in order to allow energy company, independent of its size and its consumption of irradiation capacity, the utilization of the manifold technical advantages of radiation crosslinking.

As these service irradiation centers treat the products of numerous customers, they can really exploit the capacity of powerful irradiation facilities and therefore offer radiation crosslinking under the aforementioned favourable economical conditions. At the same time these centers provide the necessary expertise in radiation technology and thus assume the role of pacemaker in this field.

Today, more than 60 customers from Germany and other European countries including Italy continuously make use of the BGS irradiation service by regular deliveries.

About 150 further companies deliver products for the radiation treatment on a less regular basis. Even most of the companies in Central Europe operating their own inhouse electron accelerators are customers of BGS, because the proper facility is not versatile enough for the range of their products which can benefit from a radiation treatment.

But it would be, at least partly, misleading to consider the potential of radiation crosslinking applications primarily under the aspect of the cost of the radiation treatment per kg of a product. The versatility of radiation crosslinking and its applicability to a far greater variety of thermoplastic polymers than the chemical crosslinking processes, open important possibilities for technical innovation in the material science (9,10). On a long-term basis, the greater potential of radiation crosslinking does not lie in the successful economic competition with established chemical crosslinking processes, but in substituting more expensive materials.

To what extent such substitutions occur, depends primarily on the material properties which can be achieved by radiation crosslinked, properly formulated, eventually alloyed and/or reinforced polymers, tailored to the requirements of specific applications.

The cost of the radiation treatment is generally not a decisive factor, if more expensive materials can be substituted by a cheaper crosslinked polymers, though ionizing radiation is and will remain an expensive form of energy.

Reference

1. M. Dole, Polym. Plast. Technol. Eng., 13, 41 (1979)

2. H. G. Scott and I. F. Humphreys, Mod. Plast., 3, 38 (1973).

3. G. Poschet, Kunststoffe, 77, 792 (1987).

4. R. F. Grossman, Radiat. Phys. Chem., 9, 659 (1977).

5. L. Spenadel, Radiat. Phys. Chem, 14, 683 (1979).

6. R. W. Waldran, H. F. McRae and J. D. Madison, Radiat. Phys. Chem., 25, 843 (1985).

7. H. Schmidt, Mashinenmarkt, 43 (20/10/87)

8. K. Ruehl, E. Scherp, Ger. Offen. DE 3.633.647 and 3.633.648 (14/4/1988).

9. L.Wiesner, Fortschritte der Verfahrenstechnik, 22D, 391 (1984).

10. L. Wiesner, Atomwirtschaft, 31, 491 (1986).

MODIFICATION OF SURFACE STATE OF POLYMERS BY GLOW DISCHARGE PLASMAS

Y.-S. Yeh and H. Yasuda

Department of Chemical Engineering
University of Missouri-Columbia
Columbia, MO 65211 USA

Abstract

Surface modification of polymers by low-temperature plasmas is reviewed. Two different types of plasma processing that can be utilized are plasma treatment and plasma polymerization. Surface modification is achieved through plasma-polymer surface interactions in the former and deposition of an ultra thin plasma polymer coating in the latter. A dramatic change in surface state of polymers is observed after this modification. By carefully controlling these plasma processes, excellent stability of polymer surface state with respect to the surrounding environmental changes can be attained. The unique state of plasma-modified surfaces plays an important role in many practical applications.

Fundamentals of low temperature plasma

Plasma can be broadly defined as a partially ionized gas consisting of electrons, ions, neutrals in ground and any higher excited states, and photons. The plasma employed for surface modification of polymers is generally created by an electric glow discharge and are sometimes termed "glow discharge plasmas" or "glow discharges". In glow discharge plasmas, the energy applied via the electric field is gained mostly by the electrons. Due to the great mass difference between electrons and other neutral particles, electrons do not lose significant amounts of kinetic energy when undergoing an elastic collision with other heavy particles. As a result, significant fraction of high-velocity electrons exists with energy high enough to trigger most chemical reactions. On the other hand, the gases in this type of plasma which are unaffected by the imposed electric field, are less energetic and remain relatively "cold" with a temperature close to the ambient. It is this characteristic that renders the glow discharge plasma a superb technique for surface modification of polymers.

Depending on the types of gas used, different reaction processes are provoked in a glow discharge plasma. From this viewpoint, glow discharge plasmas can be categorized into nonpolymer-forming plasmas and polymer-forming plasmas. Typical examples for the former are an argon or oxygen plasma, and a methane or an ethylene plasma for the latter. The respective reaction processes resulting from these two types of plasma are

F. Garbassi and E. Occhiello (eds.), High Energy Density Technologies in Materials Science, 33–45.
© 1990 *Kluwer Academic Publishers.*

generally known as "plasma treatment" and "plasma polymerization". The basics and principles of plasma treatment and plasma polymerization have been detailed in several books (1-3) and review articles (4-5).

Plasma Treatment

Plasma treatment of polymer surfaces uses reactions occurring in a nonpolymer-forming plasma to achieve a change in surface chemistry. When a polymer is brought in contact with a nonpolymer-forming plasma, two major reaction processes occur at the polymer surface. They are: (i) direct reactions of the plasma activated gases and (ii) polymer free radicals formation and subsequent reactions. Which process is predominant depends on the type of gas used for the plasma.

In a chemically nonreactive plasma using gases such as argon, helium, etc., no direct reactions between the plasma gases and surface atoms or molecules take place in the plasma state. Polymer surfaces immersed in the plasma are continuously subjected to the bombardment of energetic metastables and ions. As a result, momentum is exchanged between the impinging particles and surface atoms. This cascade momentum exchange process eventually causes surface atoms or fragments of surface molecules to dislodge, resulting in loss of substrate materials. The ablation process is also accompanied by the formation of free radicals (which results from homolytic scissions of covalent bonds) at the polymer surface. Subsequent reactions of free radicals by recombination results in surface crosslinking.

While in a chemically reactive plasma where gases such as O_2, N_2, NH_3, CF_4, etc. are used, direct chemical reactions of the plasma gases with polymer surfaces are the major event. The net result of these reactions is an incorporation of plasma activated species in the polymer surface. Ablation of substrate materials by a physical process is minimal except for a prolonged period of plasma exposure. However, in cases where the reaction products are volatile, ablation of substrate materials by a chemical process is possible. An example would be the treatment of a silicon-containing polymer (e.g. polydimethylsiloxane) by a CF_4 plasma where volatile reaction products such as silicon fluorides are produced.

The free radical formation by the bombardment of energetic metastables and ions just described, however, is confined only to the polymer surface and is not considered to be the major mechanism in nonpolymer-forming plasma. The fact that free radicals also are found in a much deeper region of a plasma-treated polymer surface suggests photons with energies in the ultraviolet range are the main source for free radical generation in this plasma type. This process also is referred to as "CASING" (crosslinking by activated species of inert gases) (6).

Plasma (Glow Discharge) Polymerization

When an organic vapor (e.g. methane, ethylene) is introduced into a glow discharge of a gas (e.g. argon) or a glow discharge of an organic vapor is created, a host of chemically reactive species is instantaneously formed. These reactive species can react with each other as well as with the solid surface exposed to the glow. As a result of these interactions, polymeric materials deposit onto surfaces exposed to the glow. The polymeric deposits which are termed "plasma polymers" or "glow discharge polymers" can be in the form of thin films, powders or oils. However, under certain glow discharge conditions, polymeric materials deposits predominantly in film form can be produced.

Because of its unique reaction schemes, plasma polymerization is highly system dependent. Parameters such as monomer flow rate, system pressure, discharge power input, frequency of discharge are considered as major variables to control the process and thus the properties of the products. Despite the variation of chemical and physical properties of plasma polymers with these variables, plasma polymers share some common characteristics that are significantly different from conventional polymers. They can be described in the following.

a) Plasma polymers are a new class of thin film materials. In principle, the film thickness can be tailored to a range of few angstroms to microns. However, the useful thickness range for practical applications of plasma polymers is often less than few hundred angstroms.

b) Plasma polymers formed under certain selected glow discharge conditions are free of pinholes and well adhered to the substrate material, particularly to polymers.

c) Plasma polymers do not have discernible repeating molecular units. In other words, plasma polymers of ethylene are not polyethylene. They are different in both chemical structure and properties.

d) Plasma polymer often contain high concentration of trapped free radicals (10^{18} to 10^{20} free spins-cm^3) in as-synthesized polymers.

e) Plasma polymers are generally under compressive stress (i.e., expansive force acting at the surface). The magnitude of the force varies with the nature of the monomer, glow discharge conditions and film thickness.

f) Plasma polymers do not contain linear long chain segments, but consist of large numbers of highly branched and highly crosslinked short segments. For this reason, plasma polymer surface are more immobile than conventional polymer surfaces.

Surface state of polymers

Classical surface chemistry considers solid surface as an immobile, rigid and equilibrium state. This assumption is valid for ceramic and metallic materials, but is rather unrealistic for polymers. Unlike ceramics and metals, polymers are mobile and can hardly attain true equilibrium at ambient temperature. Both long-range (segment diffusional) and short range (chain rotational) motions occur in the bulk phase as well as in the surface region of a polymer. Unless the polymer surfaces is highly crosslinked or highly crystallized, molecules or segments of molecule at the polymer surface will rearrange themselves in response to the neighboring phase to a more stable conformation so the interfacial free energy is minimized. The degree of freedom for these dynamic motions is influenced by factors such as polarity of molecules, rigidity of polymer chains, degree of crosslinking, degree of crystallinity, degree of hydration, steric hindrance, etc.. In addition, surface dynamic motions are time, temperature and environment dependent, like polymers in the bulk state (7).

The state of a polymer surface is characterized by the surface properties which are usually dominated by the type of surface atoms or molecules. In a strict sense, it is governed by the surface configuration of the polymer when placed in a specific surrounding medium. Surface configurations described here refer to the spatial arrangement of atoms or groups of atoms that form the outermost surface layer. This should be distinguished from the configuration of macromolecules. Any particular surface configuration is achieved as a consequence of conformational change of macromolecules.

Due to the relatively high segmental mobility, polymers often display different surface states when placed in different surrounding environments, like water or air. The surface state stability with respect to the change of surrounding conditions can be attained by (i) specific configurations of surface macromolecules or (ii) lack of rotational or diffusional mobility of surface macromolecules. Any polymer that is a rotationally symmetric macromolecule would likely show a minimum change in surface state when a different surrounding environment is imposed. Rotationally symmetric macromolecules just described refer to the macromolecule in its most stable conformation being symmetric with respect to the main backbone chain axis in a three-dimensional space. Thus, polymers such as polyethylene, polyethyleneoxide, polydimethylsiloxane, polytetrafluoroethylene, agar gel, etc. all fall into this category.

Surface characteristics of polymers are undoubtedly different from the bulk. In many instances, bulk polymer characteristics such as molecular structure, glass transition temperature, elastic modulus, etc. cannot be used to predict the surface characteristics of a polymer e.g. surface mobility. Our recent study on surface dynamics of nylon 6 and polyethyleneterephtalate films investigated by probing the hydrophobic groups implanted by a CF_4 plasma indicated surface mobility of the former polymer is significantly lower than that of the latter, despite its stiffer and bulkier macromolecule (8). This finding is quite a surprise, since one would expect a reverse result judging from the glass transition temperature of the polymer films.

A follow-up study to investigate the temperature dependence of some polymer surfaces is in progress by one of the authors (HY) (9). Preliminary results show that the relaxation of PET surfaces occur at temperatures below T_g of PET (69 °C), but ceases at about 10-15 °C. Evidence of surface molecular motions at temperatures below T_g of polymers also has been reported in the literature (7, 10).

Surface state of polymers modified by plasma treatment and plasma polymerization

Surface modification of polymers by a glow discharge plasma is achieved through plasma-polymer surface interactions and the deposition of a plasma polymer. While the former has an important role in the case of a nonpolymer-forming plasma, the latter prevails in a polymer-forming plasma. In both cases, the main effect of surface modification is usually limited only to the outermost surface layer (sub-micron level) of polymer substrate. While the state of the plasma modified surfaces can be tailored with relative ease, the bulk characteristics of the underlying

Table 1

Contact Angle of Water on Methane Plasma-Modified Contact Lenses and Silicone Rubber Sheets Followed by Various Post-treatment Procedures

Post-treatment of methane plasma coated substrate	Contact angle (θ)	
	Contact lens	Silicone rubber sheet
Uncoated	82.5	81
CH_4 plasma polymer coated (no post-treatment)	78.5	69.5
Wet air purging	77.5	63
Wet air plasma (10 sccm, 2 min) + wet air purging	26	23.5
Wet air plasma (10 sccm, 5 min) + wet air purging	31	26.5
Wet air plasma (2 sccm, 2 min) + wet air purging	29	29
Wet air plasma (2 sccm, 5 min) + wet air purging	37.5	32.5
Wet air plasma (10 sccm, 1 min) + wet air purging	35.5	24.5
Wet O_2 purging	70.5	60
O_2 plasma (10 sccm, 2 min, Power = 12.5 W)	24	22.5
Wet O_2 plasma (10 sccm, 2 min, Power = 12.5 W)	22.5	21
Immersion in water	62	38.5

* W/FM for the methane plasma polymer coating = 2.6 GJ/kg

polymer substrate often remain intact after the modification. An exception would be the case where polymers with high plasma susceptibility are employed. In such a case UV emission from plasma may penetrate into the bulk, thus affecting the bulk characteristics of the polymer substrate.

Polymer surfaces, after treatment by a nonpolymer-forming plasma, be it a chemically reactive plasma or a chemically nonreactive plasma, undergo replacement of surface atoms by plasma activated moieties along with an increase in surface concentration of crosslinks. The versatility of plasma processes in controlling the surface chemistry of the treated polymer can be visualized from the study regarding surface wettability of silicone contact lenses and silicone rubber sheets modified by different nonpolymer-forming plasmas (11). The wettability measured by the water contact angle for various plasma-modified surfaces are summarized in Table 1. Uncoated silicone rubber sheet has a water contact angle of 81°. After coating with a plasma polymer of methane at a W/FM of 2.6 GJ/kg, the water contact angle decreased to 69.5 °. It is shown in Table 1 that the effect of rendering surface wettability is more pronounced by postplasma treatment with a wet-air plasma or a wet-oxygen plasma than just purging the gas through the sample chamber. Nevertheless, water contact angle for the sample films treated by purging a wet air and a wet oxygen still decreased by 22% and 26%, respectively.

Although the modification may dramatically change the surface state, molecular dynamic motions at plasma-modified surfaces are not totally hindered by the resulting increase in degree of crosslinking. Surface state of the plasma-modified polymer still remains relatively unstable and may vary with the surrounding medium. The surface state instability is reflected in the decay of contact angle of water on plasma-modified silicone contact lenses with exposure time after plasma treatment (12).

Figure 1 shows the variation of water contact angle on the surface of plasma-modified silicone contact lenses with time. Three groups of these were evaluated based on water contact angle measurements. Six lenses were treated in each group. In group 1, the lenses were coated with an ultrathin layer (< 40 nm) of plasma polymer of methane followed by oxygen plasma treatment to render surface hydrophilicity. Groups 2 and 3 were those treated only with oxygen plasma and water vapor plasma, respectively. Even though the surfaces of oxygen plasma and water vapor plasma treated contact lens exhibit a relatively low water contact angle right after the treatment, the surface wettability is very unstable which shows a gradual decay during a four-month testing period. The decay is believed to be due to the burying of surface hydrophilic groups into the bulk phase of the silicone polymer.

Plasma polymerization, on the other hand, yields films with a variety of chemical and physical properties. These properties can be readily tailored by a control of the starting monomer and the glow discharge conditions. Due to the uniqueness of some of these properties, surface state of plasma polymers is quite

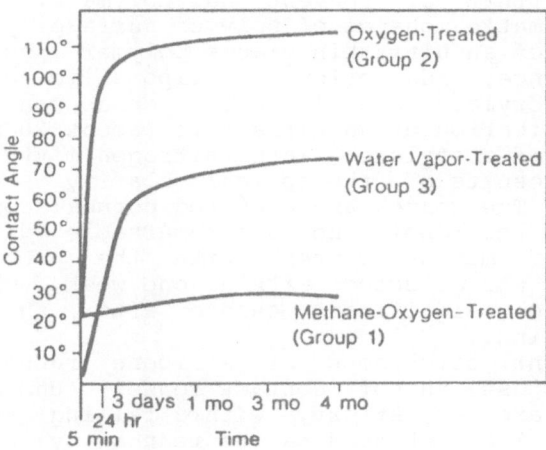

Figure 1. Water contact angle on surface of plasma-modified silicone contact lenses during four-month period. Adapted from J. E. Koziol et al., Arch. Ophthamol., 101, 1779 (1983).

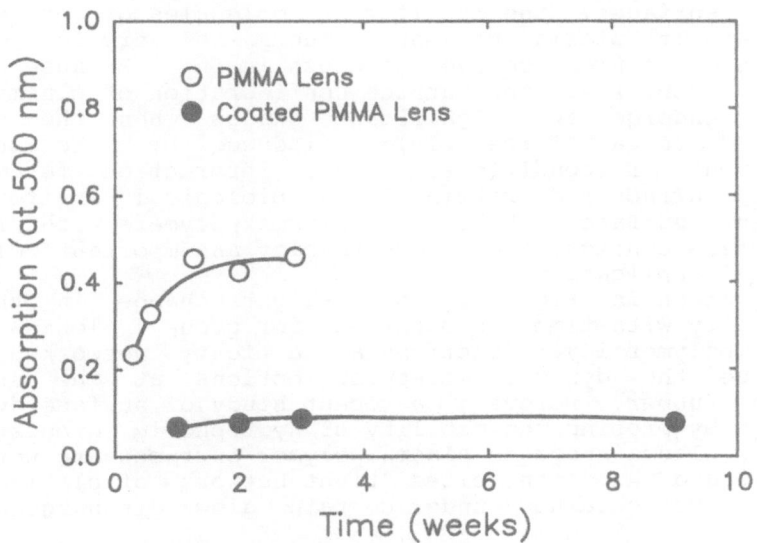

Figure 2. Increase in absorption at 500 nm (or decrease in transmittance) in the PMMA lens worn by a rabbit, due to the mucous accumulation as contrasted with the constant retention of transparency in the surface modified PMMA lens. Adapted from H. Yasuda et al., J. Biomed. Mater. Res., 9, 629 (1975).

different from that of conventional polymers. This feature is seen in the dramatic change of polymer surface properties after the deposition of an ultrathin plasma polymer layer.

For instance, adsorption of lipophilic components onto polymethylmethacrylate corneal contact lenses can be drastically reduced by application of an ultra thin (about 20 nm) hydrophilic plasma polymer of acetylene /water/nitrogen mixture (13). Figure 2 shows the results of the optical clarity study of corneal contact lenses. The coated and uncoated corneal contact lens worn continuously by the rabbit showed a remarkable difference in the accumulation of mucous matter. With the uncoated lens, the optical clarity was affected within one week, while the coated lens showed no significant change after three months of continuous wearing.

The inherent stickiness of silicone rubber poses several problems in its use as soft contact lenses. Unmodified silicone contact lenses are very sticky, with a falling angle of greater than 90° (11). The tackiness can be reduced by coating the lens with an ultrathin plasma polymer of methane. Figure 3 shows the decrease in tackiness of silicone contact lens, represented by the friction coefficient, with increasing coating thickness. The friction coefficient decreases sharply within the thickness range of a few nm to 20 nm and levels-off at a thickness of 25 nm, reaching a value of about 0.6.

As mentioned previously, most plasma polymers do share some common characteristics. Perhaps the most significant one would be the general lack of surface mobility. Compared to conventional polymer surfaces, the mobility of molecules or of segments of molecules at plasma polymer surfaces is very low due to the presence of a high degree of crosslinking. Because of the low molecular mobility, the surface configuration of plasma polymers may not undergo any significant change when the surface is subjected to a "force field" induced by the surrounding environment or conditions, e.g., interaction force between polymer surface and protein in a biological environment. The excellent surface stability of plasma polymers with respect to the surface configuration seems to play an important role in many practical applications.

As shown in Fig. 1, only slight change in the surface wettability with time was observed for group 1 lenses. the thin plasma polymer layer functions as a tight, networked layer to stabilize the dynamic molecular motions at the surface of silicone rubber. However, a recent study of surface dynamics of polymers by probing the mobility of hydrophobic groups implanted by a CF_4 plasma suggests plasma polymer surfaces are not entirely immobile and the often cited "tight network of plasma polymers" can only be obtained under certain glow discharge conditions (14).

Another practical example that makes use of the excellent surface stability of plasma polymers is seen in the study of plasma modification of vascular grafts for cardiovascular applications. We have recently carried out an extensive study (15) on the blood compatibility of polymer surfaces modified by plasma polymerization using two well established baboon models of

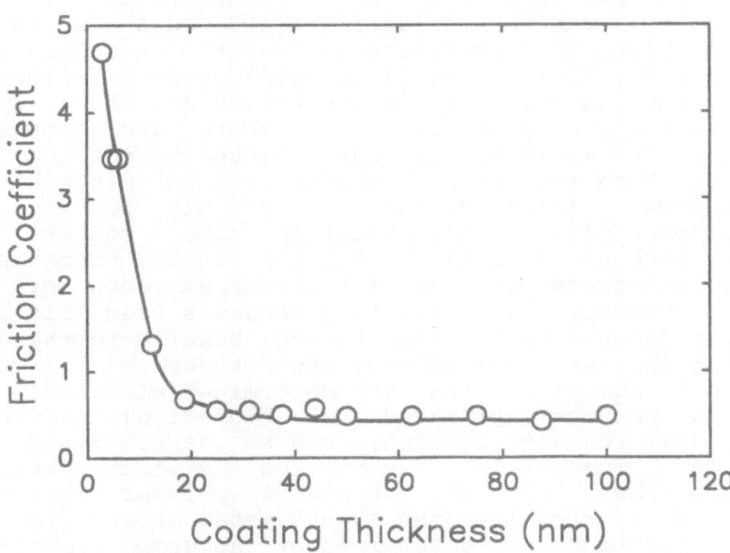

Figure 3.   Friction coefficient of contact lens coated with a plasma polymer
of methane at W/FM = 2.6 GJ/kg versus coating thickness.   Adapted
from C.-P. Ho and H. Yasuda, J. Biomed. Mater. Res., 22, 919
(1988).

Table 2

ESCA Surface Analysis of Various Plasma Polymers

Material	Elemental Ratios		
	O/C	Si/C	F/C
Plasma polymer of tetrafluoroethylene Ia	0.19	0.14	0.67
Plasma polymer of tetrafluoroethylene Ib	0.30	0.24	0.56
Plasma polymer of tetrafluoroethylene IIa	0.15	0.03	1.28
Plasma polymer of tetrafluoroethylene IIb	0.10	0.0	1.31
Plasma polymer of hexafluoroethane/H_2 I	0.33	0.18	0.03
Plasma polymer of hexafluoroethane/H_2 II	0.35	0.10	0.57
Plasma polymer of hexafluoroethane I	0.34	0.22	0.63
Plasma polymer of methane I	0.23	0.14	-
Plasma polymer of methane II	0.15	0.04	-

arterial thrombus formation (16). Smooth-walled medical grade silicone rubber tubings (Silastic) (100 cm x 3.3 mm i.d.) and expanded polytetrafluoroethylene (Gore-Tex) vascular grafts (10 cm x 4 mm i.d., 30 μm fibril length) were modified by plasma polymerization and evaluated in the baboon A-V shunt system with respect to their capacity to induce acute and chronic arterial thrombosis. The modification was achieved by coating the inner surface of both tubings and grafts with an ultrathin layer of plasma polymer (a thickness of roughly 30 nm) using a semicontinuous plasma polymerization tube coater (17). By depositing such an ultrathin layer, the surface topography of the modifying substrate can be retained, as seen by scanning electron microscopy. The starting monomers used for the plasma polymerization were tetrafluorethylene, hexafluoroethane, mixture of hydrogen and hexafluoroethane, and methane.

Table 2 summarizes the ESCA determined elemental ratios of the plasma polymers prepared for biomedical testing. These surfaces, as revealed by ESCA, can be characterized as highly fluorinated, moderately fluorinated, silicon-containing, or hydrocarbon-like. None of the plasma polymer coated Silastic tubings when inserted into the shunt system showed sign of acute platelet deposition as determined by the gamma camera imaging. the blood compatibility of these surfaces was therefore evaluated based on the chronic steady-state platelet elevation test. The rates of platelet destruction induced by these materials can be determined from these measurements. To facilitate the comparison of data obtained with respect to the exposed shunt area are summarized in Table 3.

Plasma polymers prepared in this study consumed blood platelets at a rate of 1.1 to 5.6 x 10^8 platelets per cm² of exposed surface per day. These platelets destruction rates are considered to be relatively low, as some polyurethanes, acrylic

Table 3

Platelet Consumption Rates for Various Plasma Polymers

Material	Platelet consumption (plats/cm²/day x 10^{-8})
Plasma polymer of tetrafluoroethylene Ia	1.9 ± 0.9 (5)
Plasma polymer of tetrafluoroethylene Ib	3.9 ± 1.3 (4)
Plasma polymer of tetrafluoroethylene IIa	3.7 ± 1.1 (4)
Plasma polymer of tetrafluoroethylene IIb	3.2 ± 0.8 (4)
Plasma polymer of hexafluoroethane/H_2 I	2.5 ± 1.7 (4)
Plasma polymer of hexafluoroethane/H_2 II	4.7 ± 1.2 (4)
Plasma polymer of hexafluoroethane I	3.4 ± 1.1 (5)
Plasma polymer of methane I	5.6 ± 1.8 (4)
Plasma polymer of methane II	1.1 ± 0.4 (4)

Values are mean $\pm$ 1 SEM

and methacrylic polymers and copolymers studied previously (16) with a same testing system may exceed 20 x 10⁸ platelets/cm²-day. However, the plasma polymerized surfaces are considered to be relatively non-thrombogenic regardless of the surface chemical constituents.

In contrast to the smooth-walled materials, the expanded PTFE grafts showed substantial platelet deposition over a one hour exposure. The results are depicted in Figure 4. the same graft after being coated with a plasma polymer of hexafluoroethane/H₂ showed remarkable reduction in platelet deposition compared to the untreated controls (Fig. 4). Since the surface morphology of the coated prostheses and uncoated controls

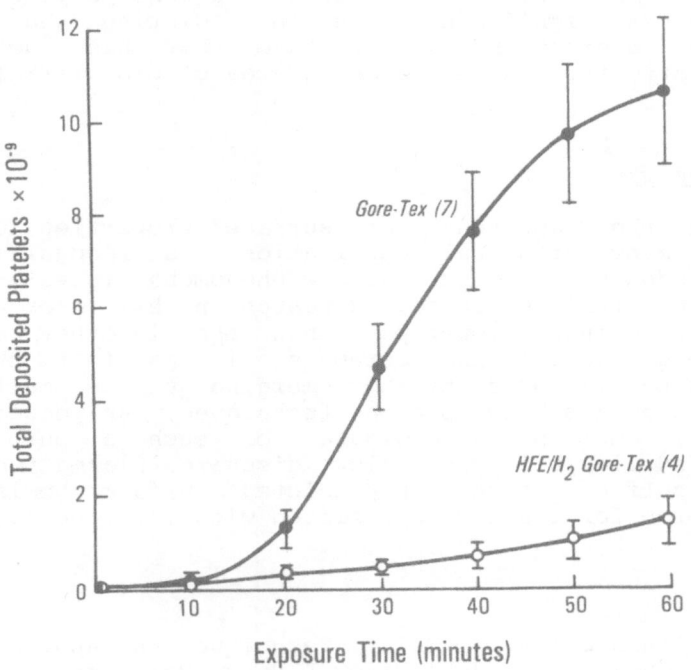

Figure 4. Platelet deposition onto untreated Gore-Tex grafts and grafts having a surface of plasma polymer of hexafluoroethane/H₂, as determined by ¹¹¹In-platelet gamma camera imaging. Values are mean ± 1 SEM. The number of studies is given in parenthesis. Animal groups studied with the treated and untreated grafts had comparable mean circulating platelets counts (466,000 ± 88,000/mℓ and 486,000 ± 70,000/mℓ, respectively) and blood flow rates (175 ± 13 mℓ/min and 163 ± 20 mℓ/min, respectively). Adapted from Y.-S. Yeh et al., J. Biomed. Mater. Res., 22, 795 (1988).

are identical as observed from the scanning electron micrographs, the results suggest the unique surface state of plasma polymers plays a major role in mediating thrombus formation. while plasma-modified expanded PTFE grafts were studied in this work, similar results were obtained with Dacron vascular grafts after coating with a plasma polymer of tetrafluoroethylene (18).

It can be concluded from this study that plasma polymerized surfaces, in both smooth and textured configurations regardless of surface composition and constituents, tend to be relatively nonthrombogenic. The nonreactive feature of plasma polymers toward blood appears to be dictated by some denominator characteristics of plasma polymers, and less governed by the starting monomers for plasma polymerization. The excellent stability of plasma polymers in terms of surface configuration is thought to be responsible for the favorable blood-surface interactions. The results of this study suggest polymer surface dynamics play a significant role in mediating the complex blood-surface interactions, a feature that has not drawn considerable attention in the research area of biomaterials.

Concluding remarks

Although the importance of surface properties of solid polymers in many practical applications is recognized, the present knowledge of polymer surface phenomena is very limited and has not been fully explored. Research in this area has been progressing at a much slower pace than that in other areas of materials, e.g. metals and ceramics. It is felt that this situation can be improved by the emerging surface modification techniques. Glow discharge plasma techniques, as described in this article, could be a candidate for such a purpose. The versatility and flexibility of glow discharge plasma techniques in tailoring surface state with predicted surface chemistry may open a new forum for research associated with interfaces.

References

1. J. R. Hollahan and A. T. Bell, "Techniques and Applications of Plasma Chemistry", John Wiley & Sons, New York, 1974

2. H. Yasuda, "Plasma Polymerization", Academic Press, Orlando, Florida, 1985

3. H. V. Boenig, "Fundamentals of Plasma Chemistry and Technology", Tehnomic Publishing Co., Lancaster, Pennsylvania, 1988

4. H. Yasuda, Plasma for Modification of Polymers, in "Plasma Chemistry of Polymers", M. Shen, Ed., Marcel Dekker Inc., New York, 1976, p. 15

45

5. W. R. Gombotz and A. S. Hoffmann, <u>Gas-Discharge Techniques for Biomaterial Modification</u>, CRC Critical Reviews in Biocompatibility, <u>4</u>, 1 (1987)

6. R. H. Hansen and H. Shonhorn, J. Polym. Sci., <u>B4</u>, 203 (1966)

7. J. D. Andrade, D. E. Gregonis and L. M. Smith, <u>Polymer Surface Dynamics</u>, in "Surface and Interfacial Aspects of Biomedical Polymers", Vol. 1, Plenum Press, New York, 1985, p. 15

8. T. Yasuda, T. Okuno, K. Yoshida and H. Yasuda, J. Polym. Sci. Phys. Ed., <u>26</u> (1988)

9. T. Yasuda and H. Yasuda, private communication

10. J. F. M. Pennings and B. Bosman, Colloid & Polym. Sci., <u>257</u>, 720 (1979)

11. C.-P. Ho and H. Yasuda, J. Biomed. Mater. REs., <u>22</u> 919 (1988)

12. J. E. Koziol, G. A. Peyman and H. Yasuda, Arch. 'Ophtalmol., <u>101</u>, 1779 (1983)

13. H. Yasuda, M. O. Bumgarner, H. C. Marsh, B. S. Yamanashi, D. P. Devito, M. L. Wolbarsht, J. W. Reed, M. Bessler, M. B. Landers, D. M. Hercules and J. Carver, J. Biomed. Mater. Res., <u>9</u>, 629 (1975)

14. T. Yasuda, K. Yoshida, T. Okuno and H. Yasuda, J. Polym. Sci. Phys. Ed., <u>26</u>, 2061 (1988)

15. Y.-S. Yeh, Y. Iriyama, Y. Matsukawa, S. R. Hanson and H. Yasuda, J. Biomed. Mater. REs., <u>22</u>, 795 (1988)

16. S. R. Hanson, L. A. Harker, B. D. Ratner and A. S. Hoffmann, J. Lab. Clin Med., <u>95</u>, 289 (1980)

17. Y. Matsuzawa and H. Yasuda, J. Appl. Polym. Sci., Appl. Polym. Symp., <u>38</u>, 65 (1984)

18. A. M. Garfinkle, A. S. Hoffman, B. D. Ratner, L. O. Reynolds and S. R. Hanson, Trans. Am. Soc. Artif. Intern. Organs, <u>30</u>, 432 (1984)

Part II

TUTORIAL LECTURES

NEW FRONTIERS IN MATERIAL PROCESSING USING THERMAL PLASMA TECHNOLOGY

Maher I. Boulos

Department of Chemical Engineering
Université de Sherbrooke
Sherbrooke, Qué, Canada, J1K 2R1

Abstract

A review is made of the fundamental aspects involved in material processing using thermal plasma technology. The description of plasma generating devices covers d.c. plasma torches, d.c. transferred arcs, radio frequency (r.f.) inductively coupled plasma torches and hybrid combinations of them. Emphasis is given to the identification of the basic energy coupling mechanism involved in each case and the principal characteristics of the flow and temperature fields in the plasma. Materials processing techniques using thermal plasmas are grouped in two broad categories depending on the role played by the plasma in the process. The simplest and most widely used processes such as spheroidization, melting, deposition and spray-coating makes use of the plasma only as a high temperature energy source. Thermal plasma technology is also used in applications involving chemical synthesis in which the plasma is used as a source of chemically active species. Examples of such applications are, the synthesis of titanium dioxide pigment, high purity synthetic silica and a large number of high purity ultrafine ceramic powders such as Al_2O_3, SiC, Si_3N_4, TiN, TiB_2.

Introduction

Plasmas are ionized gases composed of a mixture of molecules, atoms, ions and electrons in local electrical neutrality. Different types of plasma are commonly used in material processing. These can be classified in two broad groups, the equilibrium or thermal plasmas, and the non-equilibrium or cold plasmas. As shown in Fig. 1, thermal plasmas are characterized by their relatively high electron or particle densities, of the order of $10^{20}-10^{28}$ m^{-3}, and low particle energies, around 1 eV which corresponds to a particle temperature of the order of 8000 K. They are generated at atmospheric pressure or soft vacuum conditions and generally exist in local thermodynamic equilibrium in which the electron temperature is close to the heavy particle temperature ($T_e \approx T_h$.). In contrast, non-equilibrium plasmas are characterized by their relatively low electron or particle densities, less than 10^{20} m^{-3}, and their high electron temperature, of the order of a few eV. They are mostly generated under low pressure conditions, less than 1 torr, and exhibit strong deviations from kinetic equilibrium with the electron temperature considerably higher than that of the heavy particles ($T_e \gg T_h$).

F. Garbassi and E. Occhiello (eds.), High Energy Density Technologies in Materials Science, 49–64.
© 1990 Kluwer Academic Publishers.

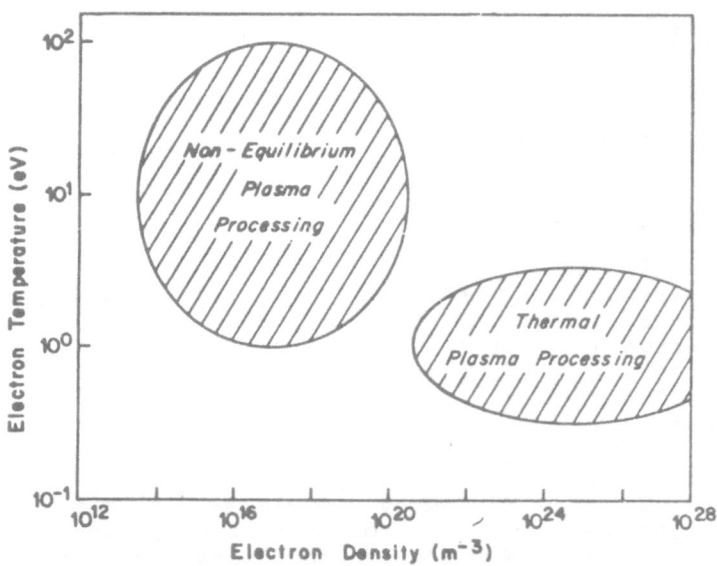

Figure 1: Typical ranges of electron temperatures and electron densities for equilibrium and non-equilibrium plasmas

In material processing, low-pressure cold plasmas are mostly used in plasma etching and deposition processes and in plasma surface modifications in general. Their effectiveness is through the reactivity of the chemically active species present rather than the total energy available in the plasma. Thermal plasmas, on the other hand, are often used in material processing for their high energy densities and for their ability to heat, melt and, in some cases, vaporize the materials to be treated. Thermal plasmas are also increasingly used as a source of reactive species at high temperature in plasma chemical synthesis of high purity materials.

The objective of the present paper is to give a general overview of thermal plasma applications in material processing. Rather than attempting to cover the subject in an exhaustive way, the review is selective and limits its coverage to some of the most recently developed applications such as plasma melting and deposition, plasma spray-coating and the plasma synthesis of ultrafine powders of high purity materials. The presentation of the paper is divided into two principal sections, the first dealing with plasma generating devices, their principal characteristics, common features and basic differences between them. This is followed by a review of specific plasma applications with emphasis given to a discussion of the fundamental aspects involved in each of these processes and to identification of potential industrial developments and pressing research needs.

Plasma generation

In this section a brief review is given of the principal plasma generating devices currently used in material processing. These include d.c. plasma torches, d.c. transferred arcs, radio frequency (r.f.) inductively coupled plasma torches and hybrid combinations of them. In each case a description is made of the basic energy coupling mechanism involved and the principal characteristics of the flow and temperature field in the plasma. It should be emphasized that different plasma generating devices tend to give rise to substantially different processing

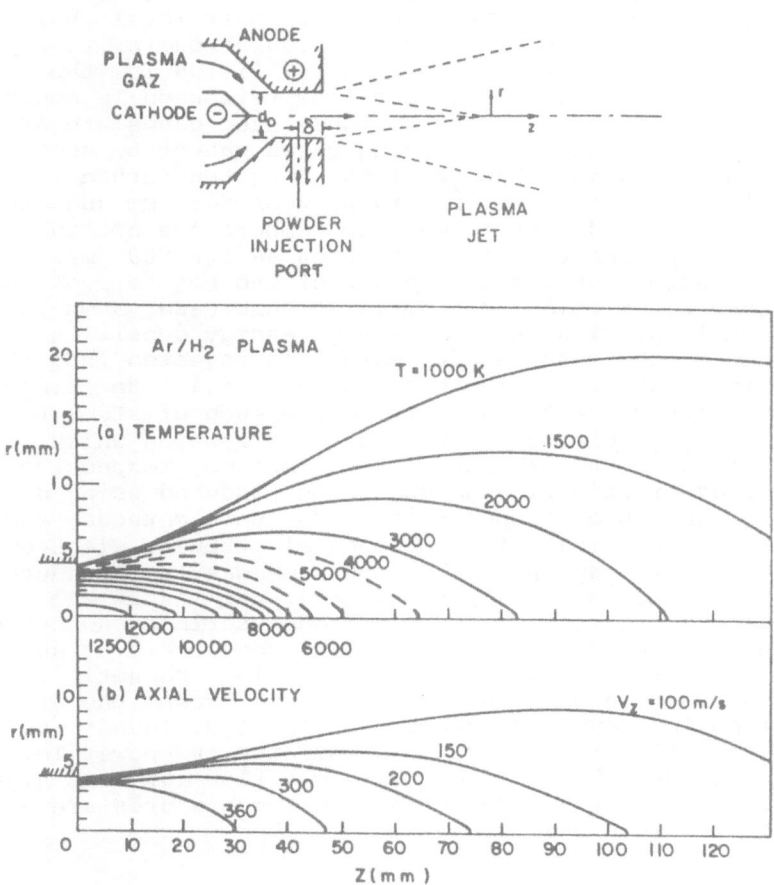

Figure 2: Schematic representation of a d.c. plasma torch with a stick cathode and the associated temperature and velocity fields for an atmospheric Ar/H₂ plasma jet.
Plasma power = 29 kW, gas flow rates = 75 L/min (Ar) + 15 L/min (H₂), d_0 = 8 mm [after Vardelle et al (1)].

conditions. The success of a given application of plasma technology in material processing depends to a large extent on our ability to properly match the process needs with the characteristics of the plasma generating device used.

D.C. plasma torches

D.C. plasma torches are among the most commonly used generating devices in material processing. These produce a high temperature plasma jet in which the material to be processed can be injected for in-flight melting and vaporization. Two principal types of d.c. plasma torches are used which differ in their electrode configuration and in the type of electrodes used.

Torches with a thoriated tungsten stick cathode and an annular copper anode are typically used at power level below 100 kW. A schematic giving the electrode configuration in this case and the associated temperature and velocity fields in the plasma jet is given in Fig.2. The total gas flow rate used is generally below 100 L/min (STP). The most commonly used gases are Ar, He, N_2, H_2 and their mixtures. Oxidizing gases cannot be used with this type of torch to avoid damage of the tungsten cathode.

Typical nozzle diameters are between 5-9 mm and the corresponding plasma jet have maximum temperature of the order of 12000 K and plasma velocities as high as 400-600 m/s. It is important to note that overall volume of the hot core of the plasma (T > 10000 K) is relatively small in this case, 5 mm in diameter and 20 mm long, with relatively high energy densities. The residence time of the gases and/or materials injected into the plasma is relatively short (of the order of 0.5 ms). The fringes of the plasma jet are characterised by the presence of steep temperature and velocity gradients in the radial direction which can reach as high as 2000-3000 K/mm and 100-200 m/s.mm, respectively.

Plasmas of oxidizing gases can be produced using d.c. plasma torches with cold copper electrodes (both cathode and anode). Such torches comprise of two co-axial tubular electrodes separated by a small gap in which the plasma gas is introduced with a strong vortex motion. This is necessary to insure the continuous motion of the arc root on the electrode surface and subsequently distribute the heat load and the electrode ware over an as large surface area of the electrode as possible. Magnetic field coils have also been used in a few designs to keep the arc root in constant motion over the electrode surface. Industrial versions of this type of torches have been operated at power levels ranging between 100 kW and 6 MW, with gas flow rates as high as 300 m^3/h (STP-Air) for a 1 MW torch. The temperatures are below 8000 K at atmospheric pressure.

D.C. transferred arcs

Transferred arcs are characterized by having a large physical separation between the electrodes which can range from a few centimetres to almost one meter long in high power industrial furnaces. This allows for the increase of the operating voltage of the furnace for the same current rating. The latter being a key parameter which has a strong influence on the energy efficiency of the arc and electrode lifetime.

A schematic of a typical transferred arc laboratory set up and the associated temperature field in the arc column is shown in Fig. 3. The top electrode which acts as cathode is made of thoriated tungsten around which the plasma gas (argon or nitrogen) is injected to protect the cathode tip from the surrounding environment. The flow of such sheath gas is responsible for the constriction of the arc column in this region with a corresponding increase of the current density and accordingly the maximum plasma temperature which can reach as high as 20000 K. The bottom electrode, the anode, often constituted of a liquid bath of the material to be melted which obviously has to be conducting. Attaching the arc to the material to be heated and molten adds to the relatively high energy efficiency of transferred arcs since the anode losses are recovered in the work piece.

Because such plasma generating devices can be operated with a minimal plasma gas flow rate for a given power level compared to d.c. torches, the overall energy density of transferred arcs tends to be high with the majority of the arc power transferred

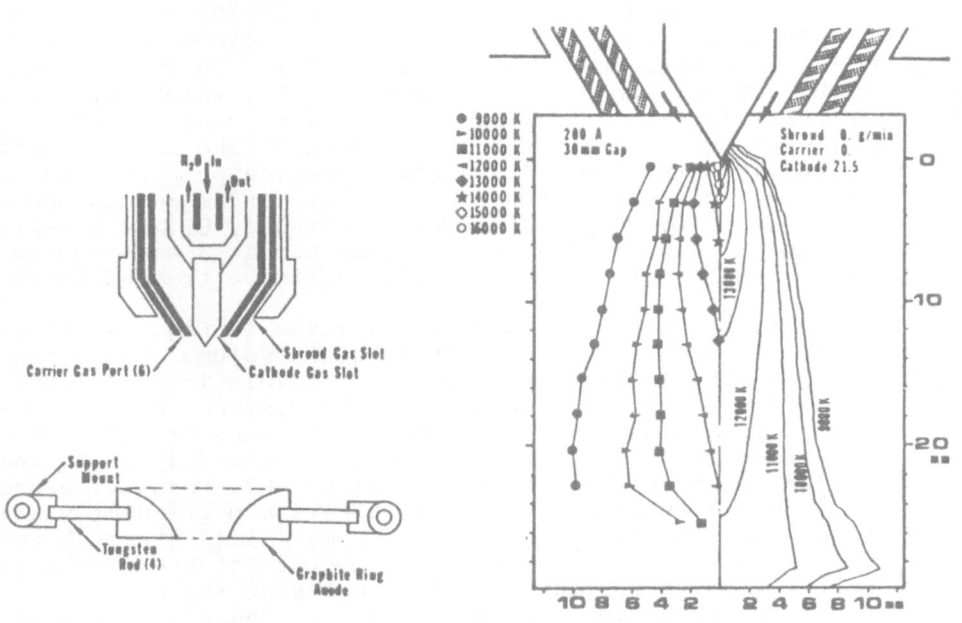

Figure 3: Schematic of a d.c. transferred arc arrangements and the associated temperature field in the arc column.
Electrode separation = 30 mm, arc current = 200 A ,[after Young et al (21)]

54

to the arc surrounding as radiative energy and as arc root losses, (Gauvin (3)). The design of transferred arc furnaces requires accordingly careful considerations in order to recover as much as possible of the radiative energy as in the case of the plasma can transferred arc furnace, (Gauvin (4)), in which the powder to be heated and molten is injected tangentially around the arc column to form a falling film of molten metal heated by the high radiative energy flux from the arc.

R.F. inductively coupled discharges

In radio frequency (r.f.) induction plasma torches, (5) energy coupling to the plasma is accomplished through the electromagnetic field of the induction coil. The plasma gas does not come in contact with electrodes, thus eliminating possible sources of contamination and allowing the operation of such plasma torches with a wide range of gases including inert, reducing, oxidizing and other corrosive atmospheres. Pure argon or mixed with other gases is still the usual choice for the plasma gas largely because of its ease of ionization. The excitation frequency is typically between 200 kHz and 40 MHz. Laboratory units run at power levels in the order of 30-50 kW while large scale industrial units have been tested at power levels up to 1 MW.

A schematic representation of a typical induction plasma torch and the associated temperature and flow fields in the discharge is given in Fig. 4. The plasma is confined in a water-cooled quartz tube with several gaseous streams introduced into the discharge. These include the sheath gas, Q_3, which serves to reduce the heat flux to the walls of the plasma confinement tube and thus protect it from damage due to overheat. The intermediate gas, Q_2, mainly serves for plasma stabilization purposes and is often introduced into the discharge with both axial and tangential (swirl) velocity components. The powder gas, Q_1, is axially injected into the center of the discharge using a water-cooled probe. It serves to introduces the material to be treated in the plasma.

From the flow and temperature fields given in Fig. 4, it is noted that the plasma is considerably large in volume than corresponding d.c. plasma jets at the same power level. The local power density is therefore lower and so is the maximum plasma temperature which is of the order of 10000 K. Because of the nature of the coupling mechanism which limits energy dissipation to the outer annular region of the discharge heated by conduction and convection from its surroundings. This offers a particularly attractive means of in-flight processing of materials through the axial injection of the material into the center of the discharge without disturbing the current carrying region of the plasma.

The corresponding flow field in the discharge, Fig.4 is, characterized by its relatively low velocities, in the range of 10-20 m/s, and the presence of a recirculation eddy in the coil region caused by electromagnetic pinch effect. This is responsible for the radially inward body force acting on the plasma at a point around the middle of the induction coil causing the observed flow reversal on the upstream end of the coil, (7). In order to have the material to be treated penetrate the plasma, it is therefore necessary either to inject them into the discharge at a high enough velocity to overcome this back flow, or to have the material feeding probe penetrate the discharge to a point around the middle of the induction coil. In either case, this results in the local cooling of the plasma at the point of injection with a slight loss of energy efficiency. The relatively large volume of the plasma and the associated long residence time of the particles in the discharge, of the order of 10-20 ms, more than compensates for such an effect and makes the induction plasma ideally suited for in-flight melting of relatively large particles of refractory metal and ceramic powders at high throughputs.

Hybrid plasma torches

Hybrid plasma torches have been developed by the superposition of more than one plasma generating devices such as in the combination of d.c. and r.f. plasma torches developed by Yoshida

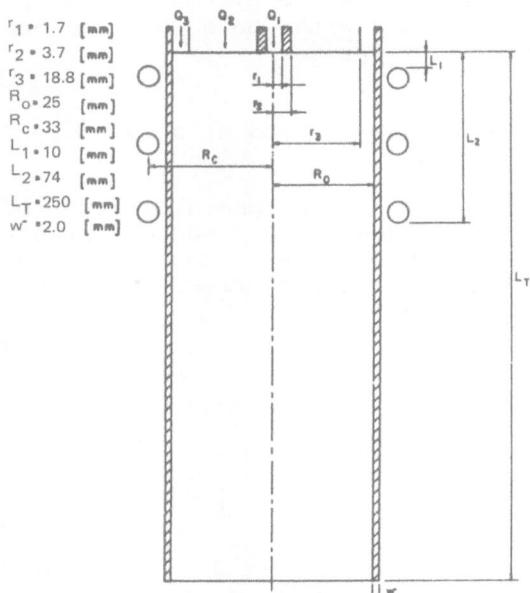
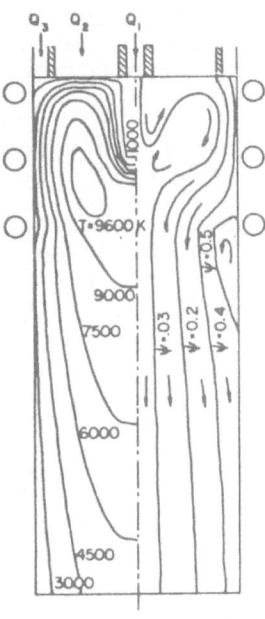

Figure 4: Schematic of an induction plasma torch and the associated temperature and flow fields in the discharge. Plasma gas: argon, Pressure: atmospheric, f = 3.0 MHz, P_0 = 3.0 kW, Q_1 = 3.0 L/min, Q_2 = 3.0 L/min, Q_3 = .4.0 L/min (STP) [after Mostaghimi et al (6)]

and Akashi (8) and the combination of two r.f. plasma torches in
tandem operating at two different frequencies, (Kameyama (9)).
Such a hybrid system is more flexible and offers a number of advantages in material processing. While they have undoubtedly
proven to be useful in specific applications involving plasma
chemical synthesis, it is unlikely because of their complexity
that their use will grow to replace alternate simpler plasma
systems.

Material processing using plasma technology

In this section a brief description is given of typical applications of thermal plasma technology in materials processing.
These can be group into two broad categories depending on the
role played by the plasma. The simplest and most widely used
technologies such as in spheroidization, melting and deposition,
and plasma spray-coating makes use of the plasma only as a high
temperature energy source. The transformations involved in the
materials in this case are mostly limited to physical changes
involving melting and rapid solidification, vaporization and
condensation. Thermal plasmas have also been used in applications
involving chemical synthesis such as in the preparation of pigments and high purity ultrafine ceramic powders such as SiC,
Si_3N_4, AlN, TiN and others. In these cases the plasma is used as
a sources of chemically active species with both physical and
chemical transformations involved in the process.

Examples of processes in which the plasma is used as a high temperature energy sources

The simplest of such processes is the spheroidization of
powder in which the material to be treated, composed of irregu-
larly shaped particles or agglomerates, are heated and molten
in-flight, and subsequently collected after freezing as they

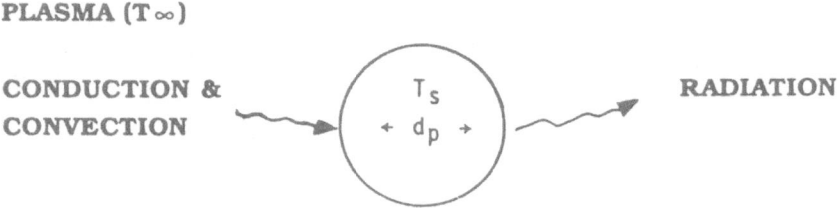

**Figure 5: Basic transfer mechanism involved in the in-flight plasma heating
and melting of particles**

emerge from the plasma. The basic heat transfer mechanism involved in the processes is schematically presented in Fig. 5.

The net energy that contributes to the heating and melting of the particles, Q_n, is the difference between the conductive and convective energy transfer from the plasma to the particle, Q_c, and the radiative loss from the surface of the particle to its surroundings, Q_r, i.e.

$$Q_n = h\, a\, (T_\infty - T_s) - \sigma\, \epsilon\, a\, (T_s^4 - T_a^4)$$

where

h   =   plasma particle heat transfer coefficient
a   =   surface area of the particle

T_∞ =   plasma temperature
T_s =   particle surface temperature
T_a =   surrounding temperature

σ =   Stephan-Boltzmann
ϵ =   Particle emissivity

For a particle to melt completely during its short residence time, τ, in the plasma, the integral of the energy received by the particle should be equal of superior to the energy required to heat the particle from its initial temperature to its melting point plus the latent head of fusion of the material of the particle.

$$\int_0^\tau Q_n \, dt > m_p C_p (T_m - T_0) + m_p H_m$$

where

m_p =   mass of particle
C_p =   specific heat of particle material
T_m =   melting point of particle material
T_0 =   initial particle temperature
H_m =   latent heat of fusion of particle material
τ =   particle residence time in the plasma

Normally a certain degree of superheat is required to insure the full melting and the decrease of the viscosity of the liquid droplet formed in order that it can assume its spherical shape under the influence of the surface tension effect. The required level of superheat for complete melting of the particle will also depend on the Biot number, Bi, which is defined as the ratio of the thermal conductivity of the plasma surrounding the particle to that of the particle materials. Porous particles and particles of low thermal conductivity materials such as oxide ceramics will tend to develop important internal temperature gradients which, depending of the heat flux from the plasma to the surface

58

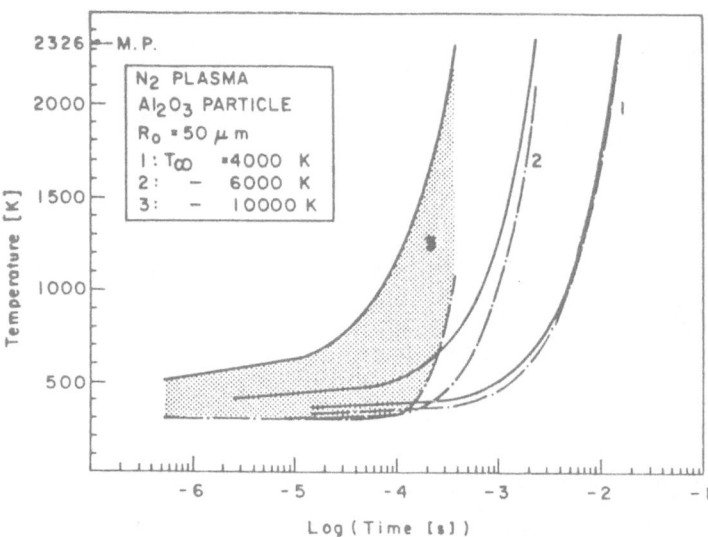

**Figure 6: Temperature history for 100-μm diameter alumina particles
immersed in nitrogen plasmas at atmospheric pressure
and different temperatures [after Bourdin et al (10)]**

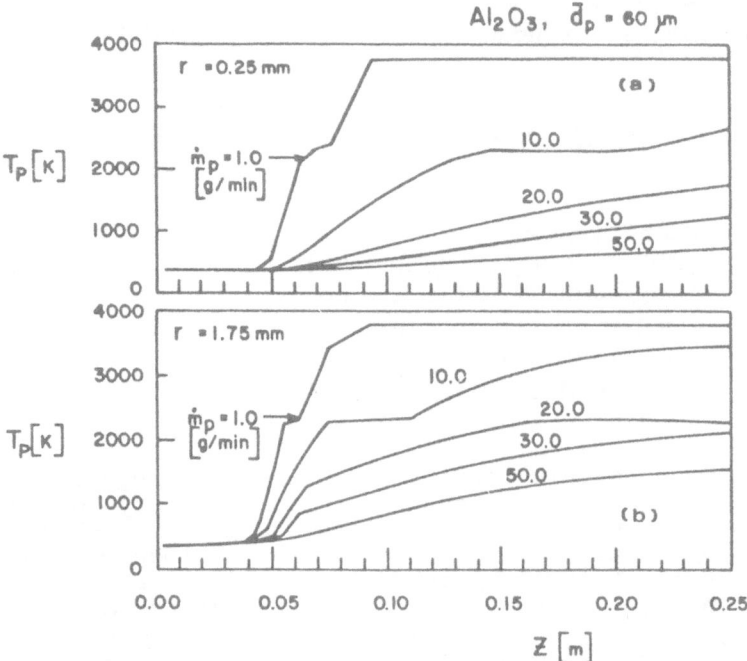

**Figure 7: Temperature history for 60-μm diameter alumina particles axially
injected into a 5 kW, atmospheric argon plasma at different powder feed
rates [after Proulx et al (11)]**

of the particle, can reach as high as 1000 K or more. Typical examples of the temperature history of 100 μm diameter alumina particles in a nitrogen plasma at different temperatures are given in Fig. 6 (after Bourdin et al (10)). The solid line in each case represents the history of the surface temperature of the particle while the dotted line gives the temperature at the center of the particle. The difference between the surface temperature of the particle and that as its center is noted to depend strongly on the heat flux to the particle as it is affected by variations of the plasma temperature.

Particle loading can also have a substantial influence on the ability to melt particles in-flight under plasma conditions. With the increase of the particle feed rate into the discharge, the energy transferred from the plasma to the particles will result in a corresponding drop in the temperature of the plasma and accordingly in the particle heating rates. Figure 7, (Prolux et al (11)) gives the temperature history of 60 μm diameter alumina particles as they travel along the axis of a 5 kW argon induction plasma at atmospheric pressure. The particle feed rate, m_P, is

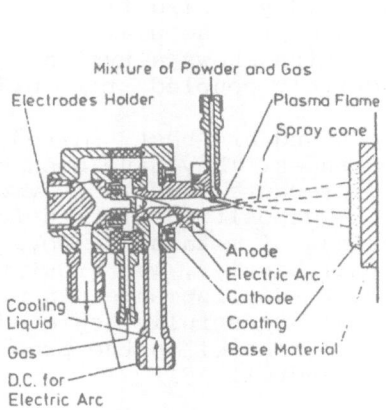

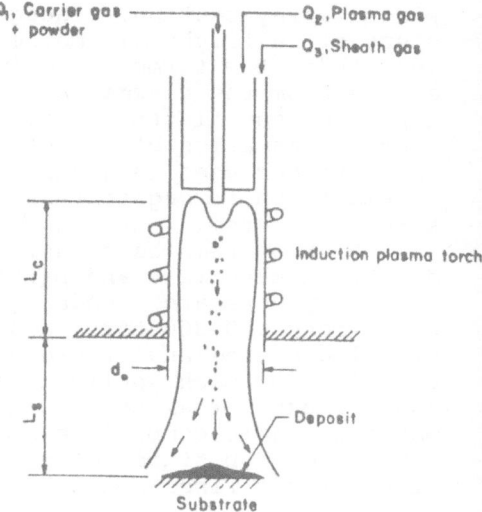

Small volume plasma
High energy density
High velocity
Gases: Ar, Ar/H$_2$, N$_2$

Large volume plasma
Low energy density
Low velocity
Gases: Ar, Ar/H$_2$, N$_2$, Air

Figure 8: Schematics of a d.c. plasma (left) and an inductively coupled r.f. plasma spraying system (right)

noted to have a strong effect on the particle temperature history and its ability to reach its melting and vaporization temperatures. In quantitative terms the results show that at the exit of the plasma, 250 mm downstream from the point of injection of the particles, all the particles would have been molten and 26 %w vaporized at a powder feed rate of only 1 g/min with the increase of powder feed rate to 10 g/min, for the given plasma power of 5 kW, as increasing weight percentage of the powder does not melt as it passes through the plasma with no particle melting achieved at powder feed rates of 50 g/min and higher.

It is important to emphasize that the powder feed rates quoted above are closely related to the plasma power level and the powder injection techniques used. The trends indicated by plasma-particle interactions and the local cooling of the plasma as a result of the passage of the particles in the discharge are important and should be given special attention in the design development of any process for the in-flight plasma treatment of powders.

While in the above analysis reference has been made to the plasma spheroidization of powders by in-flight melting and solidification, the same basic phenomena are met in the substantially more widely used process of plasma melting and deposition and plasma spray-coating. These involve essentially the same initial steps of in-flight plasma heating and melting of the powder except that the formed liquid droplets are projected, as they emerge from the plasma against the substrate on which they are deposited as flattened lamella which rapidly solidifies on impact. Schematics of two of the wide range of the plasma deposition systems used is given in Fig. 8 which shows a typical d.c. plasma torch arrangement and an inductively coupled r.f. plasma spraying system.

Both systems have their differences and are best suited for different materials and particle size ranges. They have been successfully operated under atmospheric pressure and soft vacuum conditions (50-300 torr) and used in the deposition of both free-standing bodies and of protective coatings of metals, alloys and ceramics for such applications as thermal barrier, wear resisted, and corrosion resistant coatings. Recent developments in metal matrix ceramic composites have also been successfully achieved by the plasma deposition of metals on a cold mandrel in the presence of a ceramic fiber network laid on the mandrel.

Examples of processes in which the plasma is used as a source of chemically active species

As mentioned earlier, a number of processes were developed in which the plasma is used as a source of chemically active species. These are mostly concerned with the plasma chemical synthesis of materials such as pigments, high purity synthetic silica and ultrafine high purity ceramic powders. The role of the plasma in each of these cases is in its ability to heat the reactants to

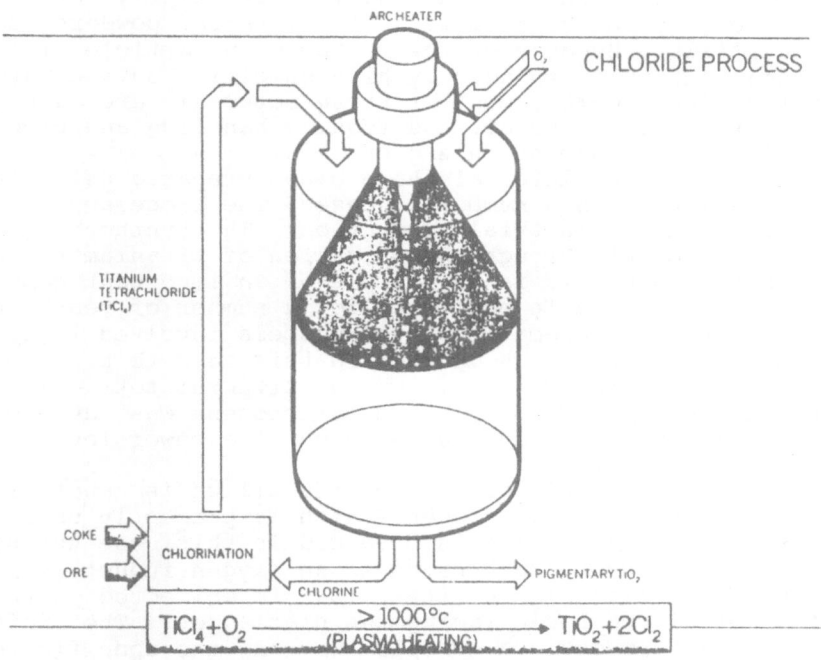

ARC HEATER

CHLORIDE PROCESS

TITANIUM
TETRACHLORIDE
(TiCl₄)

COKE

CHLORINATION

ORE

CHLORINE

PIGMENTARY TiO₂

$$TiCl_4 + O_2 \xrightarrow[\text{(PLASMA HEATING)}]{> 1000°c} TiO_2 + 2Cl_2$$

Figure 9: Schematic of the TIOXIDE titanium dioxide plasma process

relatively high temperatures under a well controlled environment in which the chemical composition of the reaction mixture can be controlled independently from its energy level.

Such reaction systems used d.c. plasma torches, transferred arcs, r.f. induction plasma torches and hybrid combinations of them. Gaseous, liquid and solid precursors have been used depending on the chemical reactions involved. Gaseous precursors are by far the most common due to their ease of handling and the control of the injection position in the reactor.

Product quench, nucleation and growth are vital to avoid the decomposition of the product through back reactions. Chemical species which are formed as stable products at high temperatures may undergo undesirable phase changes during gradual cooling to room temperature. A number of product quench techniques are currently used such as cold gas injection which can provide quench rates as high as 10^8 K/s. Quenching of the reaction products through their contact with a water-cooled surface also allows for relatively high quench rates (of the order of 10^6 K/s) with the product condensing on the cold surface. Other quench techniques have also been used such as liquid quench and fluidized bed quenching of the reaction product.

Product collection and handling is particularly critical in processes leading to the formation of ultrafine powders of high purity materials. Because of their submicron particle size range (10-20 nm) and their relatively high specific surface area (of the order of 20-80 m²/g), most of these materials are rather sensitive to exposure to the air and require handling and processing under an inert environment in a glove box.

A large range of materials have been prepared using thermal plasma technology. In a number of cases, the processes developed lead to full scale industrial production. The synthesis of titanium dioxide pigment through the oxidation of titanium tetrachloride in an oxygen plasma is an example of an important commercial success which has been in operation for a number of years by Tioxide in the U.K. A schematic of the process involved is given in Fig. 9. A d.c. plasma torch is used in this case to produce a jet of high temperature oxygen in which titanium tetrachloride is injected in the tail flame. A similar process was developed in the USSR using an induction plasma torch at a power level of 0.5 MW.

The plasma synthesis of high purity silica is another example of a successful plasma process which is presently used on an industrial scale. The silica is produced in this case through the oxidation of silicon tetrachloride in an oxygen induction plasma. The formed aerosol of high purity SiO_2 is collected on the surface of a hot silica boule facing the plasma jet. The efficiency of the process depends in a complex way on the specific reactor design used and on the surface temperature of the boule which must be maintained within close toleranceness for optimal performance.

A number of other processes involving plasma chemical synthesis have also been developed on a laboratory and pilot plant scale for the synthesis of ultrafine powders of high purity ceramic materials (Akashi (12)). These can be grouped depending on the chemical nature of the powder produced, as follows:

Oxides   : Al_2O_3, TiO_2, MgO, $Al_2O_3-SnO_2$

Carbides: SiC, TiC, B_4C, WC, TiC, B_4C, WC, TiBxCy

Nitrides: Si_3N_4, AlN, TiN, BN

Others   : TiB_2, TiCxNy and high Tc superconducting Y-Ba-Cu-oxides

These processes required the development of a large number of plasma reactors built around basically the same plasma sources, d.c. plasma jets, transferred arcs and r.f. inductively coupled discharges. they differed, however, in the way by which the reaction precursors are injected into the plasma and the product quench and collection technique used. While such studies underlined the important potential of plasma technology for the synthesis and processing of high performance materials, it also reveals important gaps in our knowledge of reaction kinetics under high temperature conditions and of the predominate nucleation and growth mechanism in the product quench and collection sections of the reactors.

Summary and conclusions

A review is presented of the fundamental principals involved in material processing using thermal plasma technology. Plasma processing of materials through in-flight melting and deposition is well established and represents a proven technology that is currently used on an industrial scale. An important surge of interest in the potential use of thermal plasmas for the synthesis of high purity materials is developing, motivated by the need to meet our continuously increasing stringent requirement in material performance. Process development has to be accompanied, however, by fundamental studies of the basic phenomena involved in momentum, heat and mass transfer under plasma conditions, plasma-particle interactions under dense loading, mass transfer and mixing patterns in a plasma reactor, high temperature reaction kinetics and nucleation and particle growth mechanism in the quench section of a plasma reactor.

Bibliography

1. M. Vardelle, A. Vardelle, P. Fauchais and M. I. Boulos, AIChEJ, 29, 236 (1983)

2. R. M. Young, Y. P. Chyou, E. Fleck and E. Pfender, "An experimental arc plasma reactor for the synthesis of refractory materials", Proc. 6th Int.Symp. on Plasma Chemistry, 1, 221 (1983)

3. W. H. Gauvin, Plasma Chem. Plasma Processing, 9, 65S (1989)

4. W. H. Gauvin and G. R. Kubanek, "A transferred arc plasma reactor for chemical and metallurgical applications", U.S. Patent No. 4-466-824 (1984)

5. M. I. Boulos, Pure Appl.Chem., 57, 1321 (1985)

6. J. Mostaghimi, P. Proulx and M. I. Boulos, Plasma Chem. Plasma Processing, 4, 129 (1984)

7. J. Chase, J. Appl. Phys., 42, 4870 (1971)

8. T. Yoshida, T. Tawi, H. Hishimura and K. Akashi, J. Appl. Phys., 54, 640 (1983)

9. T. Kameyama and K. Fukuda, "Development of an all solid state r.f.-r.f. induction plasma system, 30 kW, 1-15 MHz", National Chemical Laboratory for Industry, Tsukuba, Japan, Internal Report, Vol. 21, No. 4 (1986)

10. E. Bourdin, P. Fauchais and M. I. Boulos,
    Int. J. Heat and Mass Transfer, $\underline{26}$, 567 (1983)

11. P. Proulx, J. Mostaghimi and M. I. Boulos, Plasma Chem.
    Plasma Processing, $\underline{7}$, 29 (1987)

12. K. Akashi, Pure Appl. Chem., $\underline{57}$, 1197 (1985)

DEPOSITION AND ETCHING OF FLUOROPOLYMER FILMS BY PLASMA TECHNIQUE

Riccardo d'Agostino[1], Francesco Fracassi[1], Pietro Favia[1] and
Francesca Illuzzi[2]

(1) Department of Chemistry, University of Bari
    via Amendola 173, 70126 Bari, Italy

(2) SGS-Thomson, R&D
    via Olivetti 2, 20041 Agrate Brianza, Italy

Abstract

Deposition and etching of Plasma Polymerized Fluorinated
Monomers films have been studied in discharges fed with various
fluorinated freons and hydrogen (deposition) or with C_2F_6-O_2
mixture (etching). Both parallel plate and triode reactors with
thermally controlled substrate electrodes have been utilized in
order to ascertain the role of substrate temperature and
self-induces bias on the deposition process. This procedure
allows to give a rationale to the effects of CF_x radicals, F
atoms, positive pressure, substrate bias and temperature to the
mechanism of deposition and to the chemical structure of PPFM
films. It has also be found that both O and F atoms contribute to
the etching process though an overall first order kinetics in the
absence of substrate bias, while its superimposition induces
ion-enhanced effects.

Introduction

A great deal of work has been produced since the 70's in
plasma deposition of thin polymer films, for their technological
relevance (1, 2) and for the reliability of deposition and
etching (3) processes. With respect to this, the study of Plasma
Polymerized Fluorinated Monomers (PPMF) has recently gained much
reputation (4-9) for their application potentiality and for the
suitability of analytical techniques for studying their chemical
structure. This is the case of the popularity gained by X-ray
Photoelectron Spectroscopy (XPS) caused by the large chemical
shift induced by F in Cls spectra (5,9).
Mostly the feed utilized for the deposition of PPFM films
contain either fluorinated freons as C_nF_{2n+2}, or partially
fluorinated hydrocarbons as CHF_3, or insaturated fluorinated
compounds, as C_2F_4 or C_3F_6. Sometimes hydrogen is added in
various percentages to freons because it allows to change
continuously the concentration ratio of polymer precursors (CF_x
radicals) over etchants (F atoms). The etching process can be
studied in the same reactors utilized for deposition. Infact, it
is sufficient to introduce oxygen to the freon feed to produce
large concentration of etchant, i.e. both F and O atoms.

F. Garbassi and E. Occhiello (eds.), High Energy Density Technologies in Materials Science, 65–75.
© 1990 Kluwer Academic Publishers.

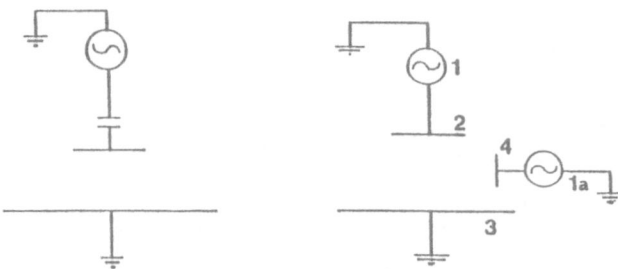

Figure 1- a) Schematics of a diode reactor.   b) Schematics of a triode reactor utilized for polymers-deposition. The target (2) and the biased deposition electrode (4) are driven by separate RF generators (1,1a). 3 is the grounded electrode.

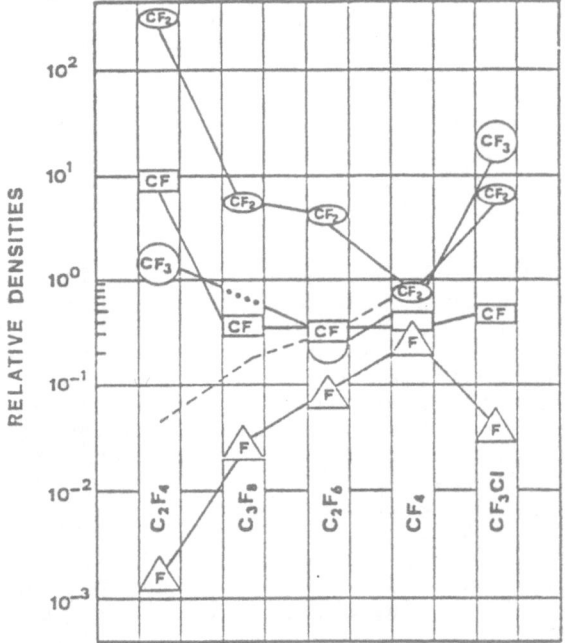

Figure 2- Hystogram of the relative densities of radicals, F atoms and electrons (dashed curve) for various fluorinated organic feeds.

Since power input, pressure, substrate temperature and bias are plasma parameters which markedly affect deposition and etching processes, their careful control and their independent variation are a need to study the kinetics of deposition and etching process. The most utilized parallel plate reactors are the diode and the triode ones, which are schematized in Fig.1.

When the diode reactor is utilized, the substrates can be set either on the grounded or on the cathodic electrode, thus allowing for a low- or high-energy positive ion-bombardment, respectively. A D.C. self-induced bias, infact, develops on the R.F. electrode in the most commonly used configuration; however, the negative D.C. bias cannot be varied without changing the power input. As a consequence, this approach can generate some misinterpretation because one cannot discriminate between the effects induced by the enhancement of the radicals and that of ions, unless an appropriate diagnostic technique is utilized. In the triode configuration the substrate electrode is smaller than the target in order to increase the self-induced and to reduce the corresponding additional power introduced in the discharge. Alternatively, both RF electrodes can be coupled to the same generator through a capacitive partitor in order to vary the self-induced bias without affecting too much the total power delivered to the discharge.

The effect of the feed

Discharges fed with different fluorinated monomers are very versatile because they can provide the plasma medium with both the polymerization precursor radicals and the etchant species. By changing continuously the radicals/atoms ratio is possible to switch from a polymerizing into an etching mode of the discharge. The species ratio can be varied either by using different pure feeds or by using additives (usually H_2 or O_2 to a Freon feed. The histogram of Figure 2 allows a comparison between CF_3Cl, CF_4, C_2F_6, C_3F_8 and the unsaturated C_2F_4. It is evident that by increasing the C/F ratio of the feed the radicals concentration increases by orders of magnitude and the atom concentration is characterized by the opposite trend. CF_3Cl escapes this simple rule because its primary step in the decomposition path is the formation of CF_3 radicals and Cl atoms, so very low densities of F atoms are allowed. C_2F_4 is practically a source of CF and CF_2 radicals. From this analysis it can be expected that CF_4 is the most etchant feed of the series and that the polymerizing ability progressively increases up to C_2F_4. Infact, CF_4 can be used as an etchant while C_2F_4 produces polymerization either over silicon or silicon dioxide.

This is however only a first-order qualitative approach since a "good " polymerizing feed must produce also high densities of charged particles to promote the formation of active sites on the polymer surface (see later). For a given freon the addition of hydrogen can significantly increase the concentration of radicals in the gas phase and change their distribution by forming progressively less fluorinated radicals. This effect can

68

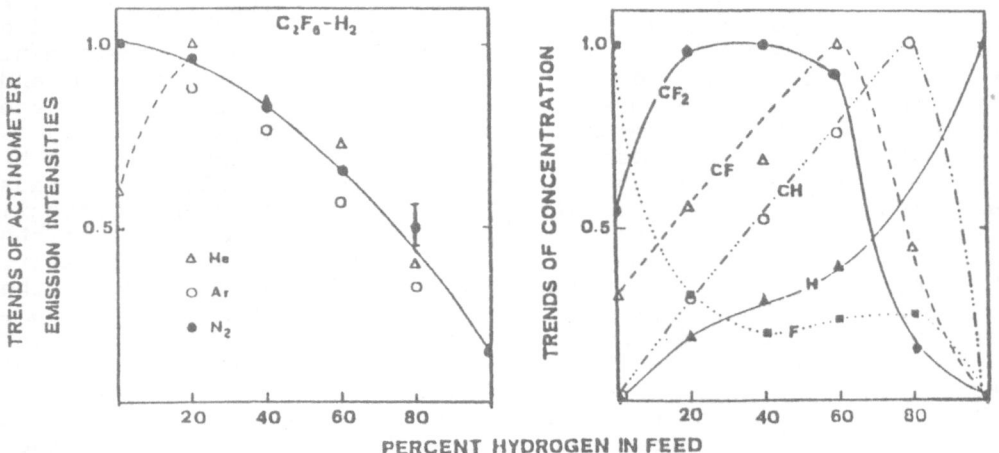

Figure 3- Trends of actinometers (He,Ar,N₂) emissions and of radicals and atoms concentrations *vs.* H₂% in C₂F₆.

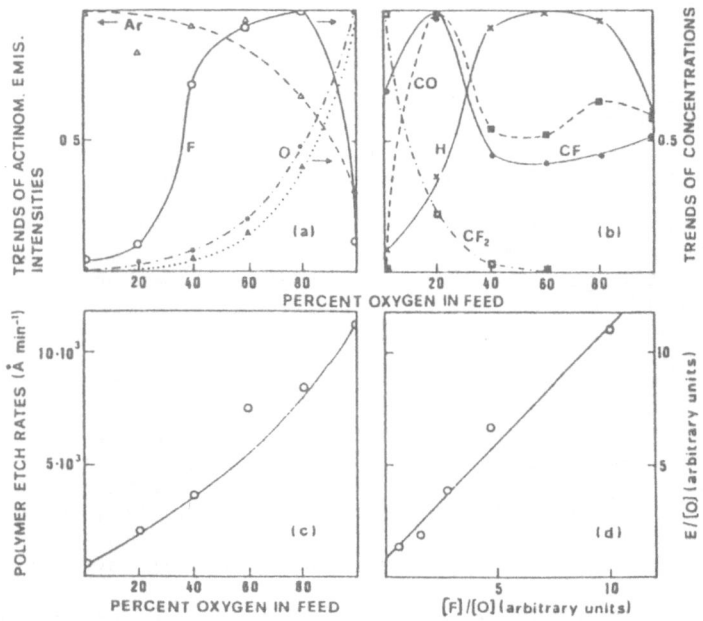

Figure 4- a) Trends of actinometer emissions and of densities of F and O atoms; b) trends of CO, CF, CF₂ concentrations; c) trends of etching rate of Low Cross Linked (LCL) films as a function of the O₂ percentage in C₂F₆/O₂ discharges performed in presence of LCL-PPFM; d) E/[O] *vs.* [F]/[O] relative values.

be appreciated in Figure 3 where the species densities are plotted as a function of the percent of hydrogen in a C_2F_6-H_2 feed.

Both Figures 2 and 3 have been obtained by Actinometric Optical Emission Spectroscopy (10). It is worth to remind the reader that, when the trends of the various actinometers (He, Ar, N_2) are coincident, as in these Figures, they represent the density of electrons, and allow to estimate the profiles of both atoms and radicals.

Figures 4a and 4b shows the profiles of actinometer emissions and those of the densities of F, O (a), of CO, CF, CF_2 (b) for C_2F_6-O_2 discharge, in the presence of PPFM films deposited in C_2F_6-H_2 discharge. The introduction of O_2 in the discharge clearly switch the plasma in an etching mode since the density of O and F atoms (etchant species) are both increased. This allows a quantitative estimation of the etching rate, as it will shown later.

The deposition process

An important condition to attain in the plasma medium to teach deposition conditions is to increase the ratio of polymer precursor with respect to etchant species i.e. [CFx]/[F]. This is not, however, the only sufficient condition to obtain high polymerization rates; two additional conditions ate required which give account of the etherogeneous nature of the processes involved in films deposition:

a - Plasma media with relatively high densities of fast electrons for substrates under <u>floating</u> conditions (rather small nega- tive self-induced bias) or of positive ions for substrates biased at negative voltages. They can ensure, with the bombardment of the solid surface a growth prooess involving the reaction of gas phase CF_x radicals with "activated" pol- ymer sites.

b - relatively low substrate temperatures because the absorption- desorption equilibrium of CF_x radicals is exothermic and can drive the overall kinetics of polymerization to a condition characterized by a negative apparent activation energy.

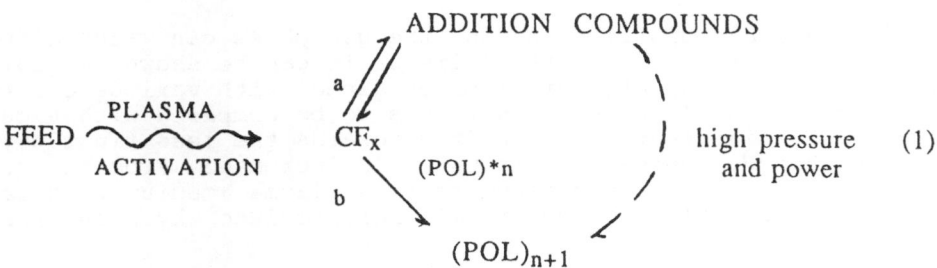

This behaviour has been interpreted as a consequence of a film growth mechanism described by the <u>Activated Growth Model (AGM)</u> (8). The radicals generated by plasma activation of the feed have a branched fate: they can either form, addition compounds or heavier radicals (branch <u>a</u>) or stick directly on polymer active sites allowing the growth of the polymeric backbone (branch <u>b</u>) In equation (1), $(POL)^*_n$ represents a polymeric chain of n C-atoms activated by charge particles bombardment, that is:

(2) $e, I^+ + (POL)_n \longrightarrow (POL)^*_n$

where the reaction (2) represents the activation step of AGM. It can be shown (9) with the use of simple mathematics that the kinetics equations (1) and (2) lead to a polymerization rate R_p, which can be expressed by:

(3) $R_p = K [CF_x] f(n_e)$

where $f(n_e)$ is a function of charged particles, either electrons or positive ions. Equation (3) has been found to fit experimental data once trends of CF_x radicals and of electrons (or positive ions bombardment for biased substrates) are obtained by AOES, provided that substrate potential and input power are not so high to trigger a sputter-etching competition of polymer films with the deposition process. This effect is clearly shown in Figures 5 (11) and 6 (12). In both figures it can be seen that, by increasing the self-induced negative bias of the substrate, the polymerization rate at first increases (due to an increased surface density of active sites) and then, after reaching a maximum, decreases due to the sputter-etching caused by momentum release of highly energetic ions impinging the surface of the polymer.

The negative effect of the substrates temperature is shown in Fig. 7, where the polymerization rate is shown to decrease with the substrate temperature (5). This behavior has been inter-preted as a consequence of the competition of the exothermic adsorption-desorption kinetics and of the chemical reaction of gas phase radicals with surface active sites (SS).

(4) $CF_{x(gas)} \longrightarrow CF_{x(surf.)}$

(5) $CF_{x(surf)} + SS \rightleftharpoons SS\text{-}CF_x$

The radicals distribution in the gas phase can also affect the chemical structure of PPFM films as it can be shown in Fig.8, where the C1s XPS spectra of films obtained with various C_2F_6-H_2 mixture are shown (5). This figure has to be compared with Figure 3, where CF_x distribution is shifted towards the less fluoricated radicals when H_2 percentage in C_2F_6 is increased. From the com-parison it can be ascertained that a plasma medium with less fluorinated radicals produces films characterized by high cross-

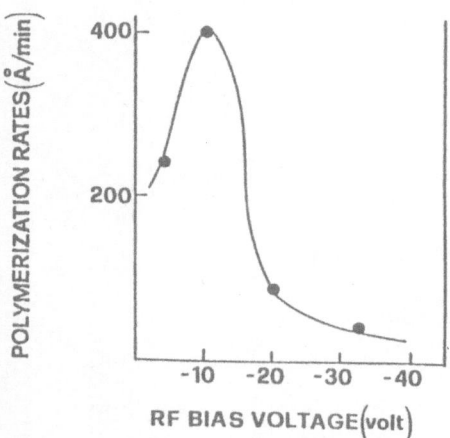

Figure 5- Effect of the imposed RF bias on the polymerization rate.

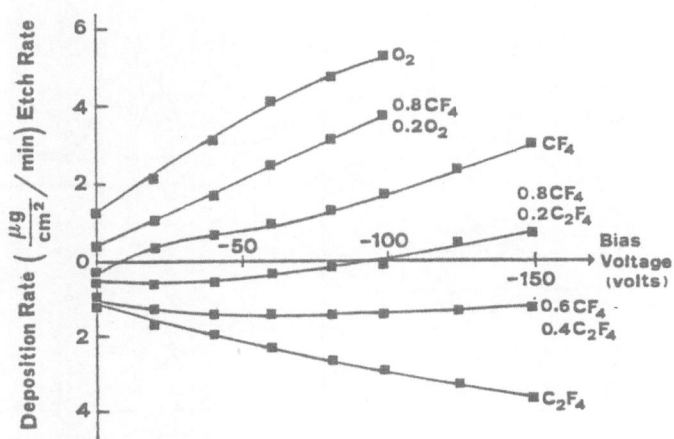

Figure 6.- Deposition and etching rates vs. bias voltage for different CF_4/C_2F_4 mixtures.

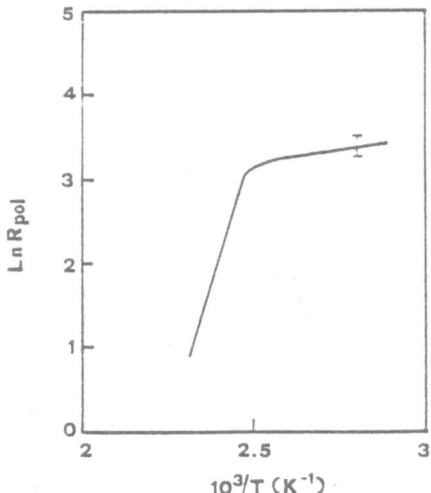

Figure 7.- Effect of the increasing temperature on the polymerization rate.

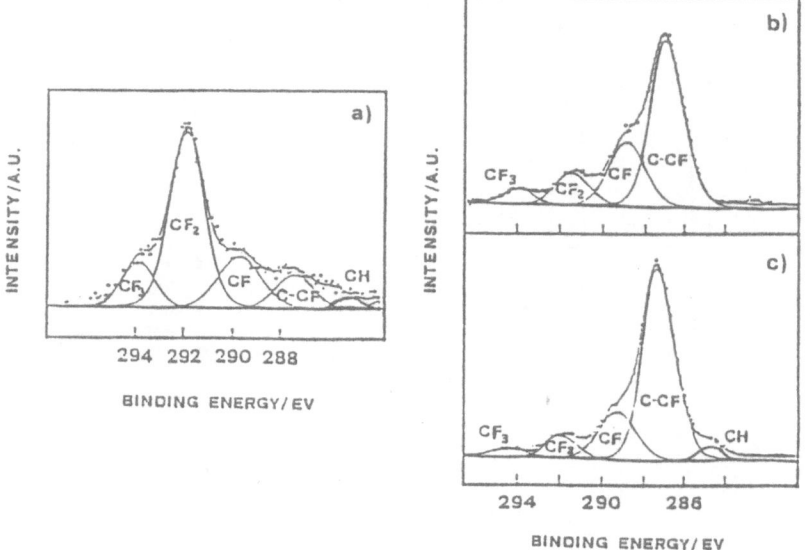

Figure 8- XPS C1s spectra of PPFM deposited at various $H_2\%$ in C_2F_6:
a) 20%; b) 50%; c) 70%.

linking and low densities of $-CF_3$ and $-CF_2$ groups. The same effect can also be produced by increasing the bombardment of positive ions.

The etching process

In Figure 4 it is clearly shown that both F atoms and O atoms are PPFM films etchants; this is an important difference between fluorinated and non fluorinated plasma polymers. Egitto et al. have demonstrated that F atoms can not be considered etchant species of non fluorinated polymers, eventhough they significantly catalyze the etching reaction of oxygen atoms. A close inspection of Figure 4 reveals that F atoms etch the polymers without any doubt since one as a definite etch rate or the polymer even when there is no oxygen added to C_2F_6 feed. The etch products which can be detected in C_2F_6-O_2 mixtures can be identified in Fig. 4b as H, CF,CO.

The data of Figure 4 allowed us to conclude that the etching process, in the absence of ion bombardment, occurs through two contemporaneous first-order kinetics depending on [F] and [O]:

(6) $E = K_F[F] + K_O[O]$

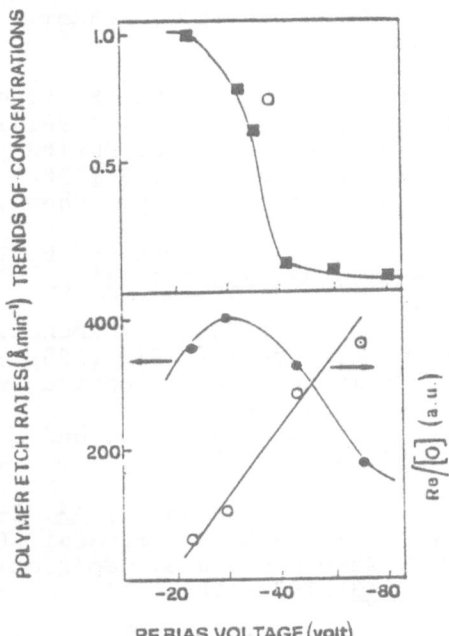

Figure 9- a) Trend of O atoms concentrations *vs.* $O_2\%$ in the feed; b) Etching rate and $R_e/[O]$ vs. imposed RF bias.

Equation 6 can be rearranged into:

(7) $E/[O] = K_F[F]/[O] + K_O$

which can be linearized by plotting $E/[O]$ <u>vs.</u> $[F]/[O]$, as shown in Fig. 4b.

The effect of bias superimposition (and of positive ion bombardment) is clearly shown in Fig.9 for a pure oxygen feed. It can be appreciate that in this case $E/[O]$ is not a function of K_O, because it increases continuously with the negative bias voltage. Is then possible to conclude that under ion bombardment an ion-induced etching process should be operative, according to:

(8) $E = K'_O[O]g(I^+)$

From this equation it can be seen that $E/[O]$ is a function of $K'_O g(I^+)$, <u>i.e.</u>of the energy and/or flux of the ions bombarding the surface. Similar results have been obtained with feeds containing both F and O atoms.

References

1. H.K.Yasuda, J. Polym. Sci. Macromol. Rev., <u>18</u>, 199 (1981) and references therein; "Plasma Deposition and treatment of Polymer Films", R. d'Agostino Ed:, Academic Press, Boston, to be published.

2a. J. Tamano, S. Hattori, S. Morita and K. Yineda, Plasma Chem. Plasma Process., <u>1</u>, 261 (1981);
2b M. Hori, T. Miwa, S. Hattori and S. Morita Plasma Chem. Plasma Process., <u>4</u>, 129 (1984;
2c S. Morita and S. Hattori, Pure Appl. Chem., <u>157</u>, 1277 (1985)

3. F.D. Egitto, F. Emmi, R.S. Horwarth and V. Vukanovic, J. Vac. Sci. Technol., <u>B3</u>, 893 (1985)

4a. R. d'Agostino, F. Cramarossa and S. DeBenedictis, Plasma Chem. Plasma Process., <u>2</u>, 213 (1982);
4b R. d'Agostino F. Cramarossa, V. Colaprico and R: d'Ettole, J. Appl. Phys., <u>54</u>, 1284 (1983);
4c R. d'Agostino, P. Capezzuto, G. Bruno and F. Cramarossa, Pure Appl.Chem., <u>57</u>, 1287 (1985).

5a. R. d'Agostino, J. Vac. Sci. Technol., <u>A3</u>, 2627 (1985)
5b R. d'Agostino, F. Cramarossa, F. Fracassi, E. Desimoni, L. Sabbatini, P.G. Zambonin and G. Caporiccio, Thin Solid Films, <u>143</u>, 163 (1986).

6. E. Kay, J. Coburn and A. Dilks in "Topics in Current Chemistry: Plasma Chemistry III", S. Veprek and M. Venugopalan Eda., Springer, Berlin, 1980.

7. N. Morosoff, H.K. Yasuda, E.S. Brandt and C.N. Reilly,
    J. Appl. Polym. Sci., 23, 1003 (1979); 23, 3449 (1979);
    23, 3471 (1979).

8. R. d'Agostino, F. Cramarossa and F. Illuzzi,
    J. Appl. Phys., 61, 2754 (1987)

9. R. d'Agostino, F. Cramarossa, F. Fracassi and F. Illuzzi
    in "Plasma Deposition and Treatment of Polymer Films",
    R. d'Agostino Ed., Academic Press, Boston, to be published.

10a.R. d' Agostino, F. Cramarossa, S. DeBenedictis and
    G. Ferraro, J. Appl. Phys., 52, 1259 (1981)
10b.R. d' Agostino, F. Cramarossa, S. DeBenedictis and
    F. Fracassi, Plasma Chem. Plasma Process., 4, 163 (1984)
10c.R. d' Agostino, F. Cramarossa, S. DeBenedictis and
    F. Fracassi, L. Laska and K. Masek,
    Plasma Chem. Plasma Process., 5, 239 (1985).

11. Unpublished results, this laboratory.

12. F. Fracassi, J. Coburn, IBM, San Josè, CA
    (private communication)

AMORPHOUS CARBON FILMS

Ludvik Martinu

Departement de Génie Physique et Groupe de Couche Minces (GCM)
Ecole Polytechnique
CP 6079, Succursale "A", Montreal
Quebec H3C 3A7, Canada

Abstract

Hydrogenated amorphous carbon (a-C:H) films possessing diamond-like properties are surveyed and discussed with respect to deposition techniques, film characteristics and proposed applications. Recent trend to further modify these films by inclusion of metals into the carbon matrices are summarized and illustrated by our latest results on a-C:H structures containing gold, silver or copper clusters.

Introduction

Study of carbonaceous thin films has witnessed remarkable progress, especially within the last ten to fifteen years, and numerous review (1-4) already exist on this subject. In the present paper we present a summary of data on amorphous hydrogenated carbon (a-C:H) structures spread across the literature; in particular, we emphasise some common feature of films deposited under apparently different conditions (first part), and we point out the attractive possibilities to further modify carbonaceous layer by incorporation of metal clusters (second part) (5).

General considerations

The interest in carbonaceous deposits of unusual characteristics appeared to arise in two steps. The first one started in the fifties with the announcement of carbon layers possessing "diamond-like properties by Schmellenmeier in 1955 (6), followed by further works on films grown on ion or electron irradiated surfaces and investigated for their electrical behaviour (e.g.7): More intensive studies on a-C:H films have been appearing since the seventies, when the interest in the basic structure as well as potential applications were widely recognized.

In most cases a-C:H films are grown in hydrocarbon (H-C) plasmas under energetic ion bombardment (therefore the nomenclature "i-C films" is sometimes used). As some microscopic characteristics of the deposits are reminiscent of diamond, the description diamond-like carbon (DLC) is often applied. These

77

F. Garbassi and E. Occhiello (eds.), High Energy Density Technologies in Materials Science, 77–87.
© 1990 Kluwer Academic Publishers.

films are basically amorphous, contrary to semicrystalline or crystalline diamond structures obtained under rather different conditions (8,9).

Some recent parallel studies on bulk carbon are worth mentioning. It has been widely accepted for a long period of time that carbon exists only in the graphite and diamond phases based on sp^2 and sp^3 hybridations. Nevertheless, a new phase diagram for carbon has been suggested by Whittaker (19) introducing a low pressure-high temperature phase of carbines (sp^1). They appear predominantly in linear chain configuration (polyene), rather than in crystalline networks. Up to now about 10 linear polymorphs have already been discovered (11). Even if their thermodynamic stability is substantially lower than that of graphite, it seems that carbines may be "frozen-in" by rapid cooling and then exist metastabilly under normal conditions.

Growth of a-C:H films

Deposition of a-C:H films is usually performed in plasma systems producing sufficient concentrations of energetic species bombarding the growing organic layers. In most cases the films are prepared from hydrocarbon monomers, often mixed with inert gases, especially argon. Radio frequency (RF) discharges have been used most often (3, 12-14). A simple version of such a system is illustrated in Fig.1. The a-C:H films are grown on the RF powered electrode, which is capacitively coupled to the RF power supply, and where a high negative DC self-bias potential V_B develops (15, 16). The growing films are exposed to the plasma and they are modified by simultaneous ion bombardment. Optical emission spectroscopy (OES) and mass spectrometry have been successfully used to study relationships between the plasma parameters and film characteristics (17). It has been shown that ions (e.g. $C_6H_6^+$ in the case of an RF discharge in C_6H_6) are extracted from the plasma glow, are accelerated by the sheath potential, and are responsible for further fragmentation on the surface. The fragments, E.G. C_2 but especially CH, are independent of the kind of hydrocarbon gas used as feed. The OES measurements are also very useful when controlling the inclusion of other components, such as metals, into a-C:H matrices (18).

Other techniques have also been applied for deposition of a-C:H films, for example DC glows discharges, where a grid is often placed in front of the cathode to supply secondary electrons so as to compensate for positive charging of the deposit (19); sputtering of carbon using unbalanced DC-magnetron (20), or large area microwave plasma deposition (21), laser evaporation of carbon combined with Ar^+ ion beam bombardment of the condensing layer (23), and film growth from H-C gases introduced directly into a saddle field source (24) have also been reported.

Comparing the published data, deposition parameters applied in RF plasma systems are usually in the following ranges: power 50-300 W, pressure 1-10 Pa, flow rate 1-10 cm^3/min(STP) and substrate bias V_B from -150 to -1000 V. In DC systems the power is

Fig. 1:

Schematic arrangement of a
system for deposition of a-C:H
films: S - substrate position,
VF - to the RF power supply,
EB-evaporation boat, P-pumps,
W - quartz window, G - vacuum
gauge, Sh - shutter.

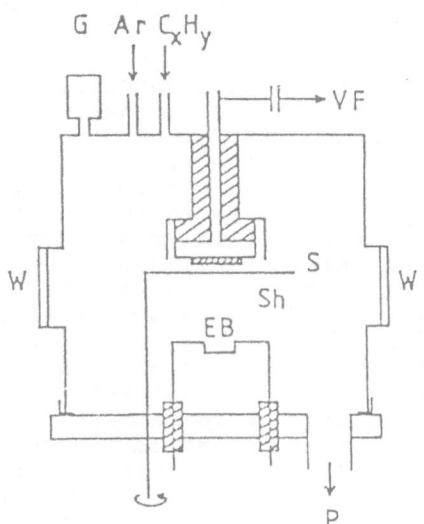

Table 1.   Basic characteristics of a-C:H films

Structure: Multiphase system: amorphous diamond-like amorphous graphitic amorphous polymeric (voids) Coordination: fourfold (sp³) - 2/3 of C atoms threefold (sp²) - 1/3 of C atoms
Hydrogen concentration: 20 - 50 at.%, 1/3 - 1/2 not bonded to carbon
Density: 1.5 - 2.0 g/cm³ (higher conc. of H) 2.0 - 3.4 g/cm³ (lower conc. of H) compared to: soft plasma polymers: 1.0 - 1.2 g/cm³ glassy carbon: 1.50 - 1.55 g/cm³ graphite: 2.3 g/cm³ diamond: 3.5 g/cm³
Optical properties: refractive index: 2.0 - 2.9 optical gap: 1.0 - 2.0 eV
Electrical properties: conductivity: 10^{-10} - 10^{0} $(\Omega cm)^{-1}$ compared to: graphite: 10^{2} $(\Omega cm)^{-1}$ diamond: 10^{-18} $(\Omega cm)^{-1}$
Microhardness: 10 - 50 GPa compared to: sapphire: 24 GPa diamond: 70 GPa

usually between 100 and 400 W, pressure 0.1-1.0 Pa and flow rate
1-5 cm³/min(STP). In most cases the resulting deposition rates
are between 0.1 and 1.5 nm/s, for good quality films.

Since film deposition usually occurs under intense ion bombardment, the mechanism responsible for the formation of the
amorphous random structure of a-C:H layers is thought to proceed
via one of the two following ways: a) dissipation of the energy
of impinging particles by rapidly collapsing (10^{-11}s) energy
spikes leading to metal stable phases and "frozen-in" states; or
b) preferential resputtering of weakly bonded surface atoms, accompanied by build up of those atoms having the strongest bonds.

When increasing the total energy delivered to the growing
surface, the film character shifts from polymer-like, soft deposits to DLC films; at very high energies graphitization is observed. This means that only a certain range of energies give
rise to the DLC structure, characterized by a high ratio of the
sp^3/sp^2 hybridizations present in the microstructure, and containing substantial concentrations of hydrogen.

It also follows that the most important parameter is the net
energy flux delivered to the growing surface. That is, the key
factor is not only the energy of the impinging particles, but
also their current density, and the duration, for which the growing surface is exposed to this flux. Therefore, high quality
films are usually prepared at relatively low deposition rates
(for example, using higher substrate temperatures or mixing an
inert gas with the hydrocarbon).

Properties of a-C-:H films

The properly adjusted energy flux during film growth gives
rise to properties such as high density, hardness, chemical in-
ertness and low coefficient of friction. Typical ranges of values
and parameters based on our own measurements, and literature data
are summarized in Table 1.

It as been generally accepted that the unique properties of
a-C:H films arise principally from high concentrations of tetrahedral (sp^3 coordination) C-C bonds. These films also contain a
considerable amount of hydrogen (see Tab.1), part of which is
not chemically bonded, and which is linked with the films' observed high compressive stress (12). The a-C:H films are thus
considered multi phase mixtures of short range order components:
"diamond-like" (sp^3), graphitic (sp^2), and polymeric (sp^1). The
resulting characteristics are then governed by the prevailing
phase. As an example, high temperature annealing induces a
decrease in the sp^3 and sp^1 components, and an increase of graphitic portion, accompanied by the appearance of voids (25). The
predominant diamond-like component has been suggested to consist
of puckered n-fold (n=3-8) rings, interconnected by strong
crosslinks of tetrahedral bonds (26).

Recently, a-C:H films have been shown to possess higher atom
number density than any other known hydrocarbon (27). Films with
hydrogen concentrations between 50 and 60 at.% adopt a structure
with an average coordination number close to the theoretical val-

Table 2. Variations of densities of carbonaceous films
         (after ref. 4)

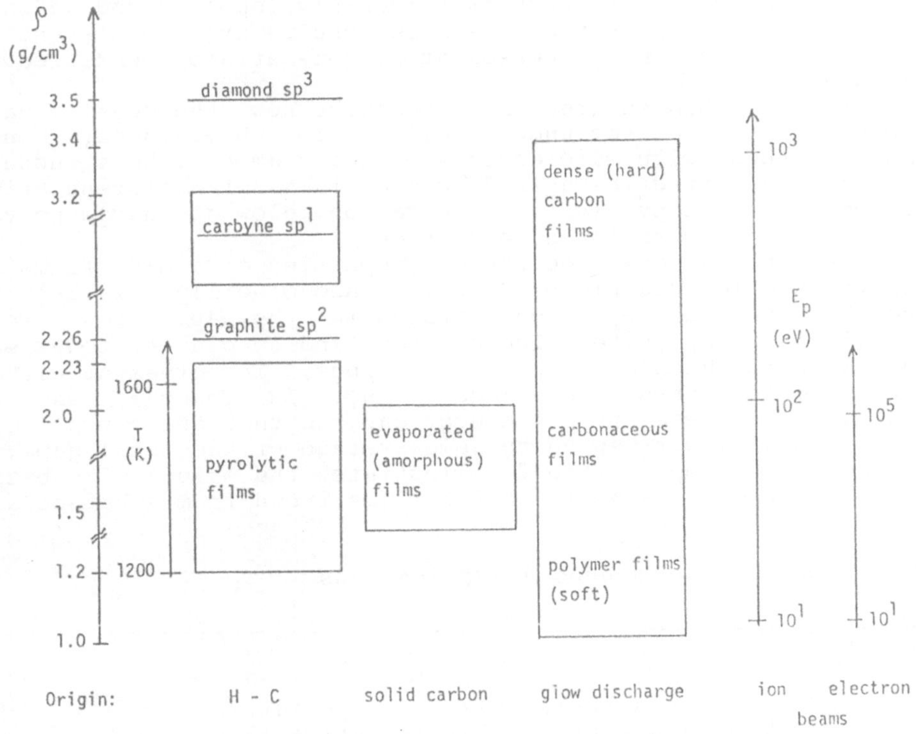

Origin: H – C   solid carbon   glow discharge   ion   electron
                                                                          beams

Fig. 2:

Optical gap as a function of
the negative bias voltage V_B
for a-C:H films deposited
from a benzene(40%)/argon(60%)
mixture at the total pressure
of 3 Pa.

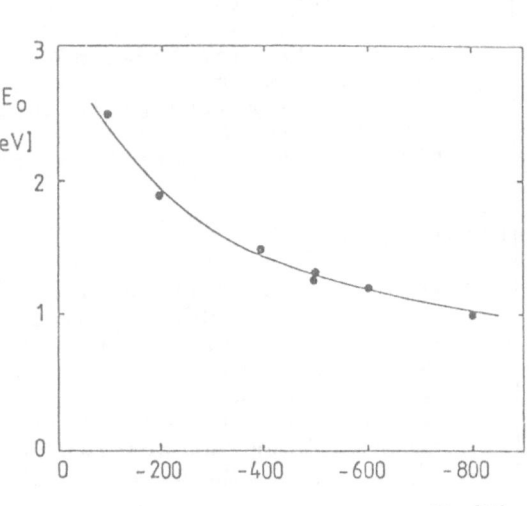

82

ue at which stabilization by bonding and destabilization by
strain energy are balanced (28). This coordination number is
achieved by the corporation of the hydrogen, providing strain
relief, and by the presence of trigonally (sp²) bonded carbon in
the carbon skeletal network. These results are consistent with
the non equilibrium processes at plasma-surface boundaries discussed above.

It is then interesting to observe how the density varies
between films deposited under "mild" (low bombardment) conditions, and those with more densely packed diamond-like structures.
This is illustrated in Table 2 where films of different origins
are compared: H-C pyrolysis, evaporation, glow discharge polymerization, and beam assisted deposition.

The optical and electronic properties of a-C:H films have
been treated for example by Robertson and O'Reilly (29,30). They
have shown that all electronic states in the gap region are due
to orbitals on sp² sites. These sites tend to cluster together in
aromatic units such as fused 6-fold rings, in agreement with the
above described structural model (26). In general, each such
cluster can possess its own energy gap, so that the overall optical gap E_0 will correspond to some average of the local gaps.

As an example, Fig.2 illustrates the effect of substrate
bias V_B on the E_0 value for a-C:H films frown from a butane/argon

Table 3. Summary of proposed applications

a-C:H films:
-corrosion and abrasion resistant coatings
-antireflection coatings for IR optics (e.g. for Ge at 10.6 µm) and solar cells
-tribological applications, protection against severe environments (sand, salt water)
-dry lubricants
-hermetic barrier coatings on optical fibers and on suboptical memory discs to give longer life times
-laser writing (change in conductivity upon irradiation from 10^{-6} to 10^{+1} Ωcm)
-biocompatible material for protection, and for promoting cell growth

metal/a-C:H films:
-graded index, multilayer structure for solar energy conversion (Cr,Mo, stainless steel)
-adhesive and abrasive wear with low friction (W,Ti,Ta,Ni,Cr)
-optical recording
-coatings for electrical sliding contacts (Ta,Sn,Rn,W,Ti)
-anticorrosion and decorative black coatings (Ti)

mixture in an RF discharge system like that in Fig.1. Hard DLC films were encountered at V_B exceeding -200 V; the E_0 values, derived from Tauc plots which are usually obeyed for a-C:H (25,30), are seen to decrease with rising sp^2 content.

There have been some attempts to modify the films by incorporation of other elements: for example, addition of about 1% of N,P or I has been shown to increase the film conductivity by two orders of magnitude (31). Some effort has been devoted to obtaining hard carbon films from fluorocarbon monomers, and to controlling optical absorption in the visible region (3,32). It has been reported that above a certain concentration of fluorine the deposition rate increases rapidly, but the films degenerate to soft fluorocarbon plasma polymers.

On account of their unique characteristics, a-C:H films have been proposed for a number of applications, examples of which are listed in Table 3.

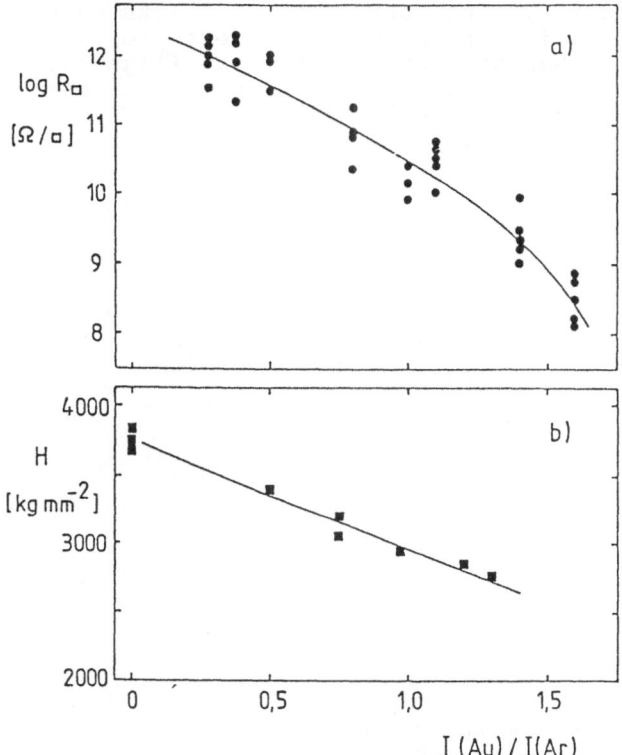

Fig. 3:

Sheet resistance $R_\square$ (a) and microhardness H (b) versus the optical emission intensity ratio I(Au,267.6nm)/I(Ar,418.0nm), which characterizes the increasing amount of gold in the a-C:H matrix deposited from a butane (70%) + argon (30%) mixture at 50 W power, V_B = -500 V and 3 Pa total pressure.

Metal filled a-C:H matrices

Deposition and properties of metal-containing a-C:H films have recently been reviewed in ref. 5. If the metals are to be incorporated simultaneously during film growth in an RF glow discharge, they are usually evaporated from a separate source, as depicted in Fig.1: The composition can easily be controlled by OES (18), so as to adjust and maintain the desired ratios of species in the gas phase and hence in the resulting layers.

When a metal is evaporated into a hydrocarbon plasma with a negatively biased substrate, the condensation processes at the surface are strongly affected by ion bombardment as discussed in a previous section. The ions carry additional energy, enhancing metal nucleation; a-C:H/metal composites display better structural stability than metal filled plasma polymers on account of their more rigid dielectric matrix (5,33). When reactive metals are incorporated, a three phase system may be formed, comprising the a-C:H matrix, the pure metal and its carbide (5,34). The embedded metal is expected to affect mostly the electrical. optical and mechanical properties, as illustrated by our recent results below. Numerous applications have been proposed, and these are listed in Table 3.

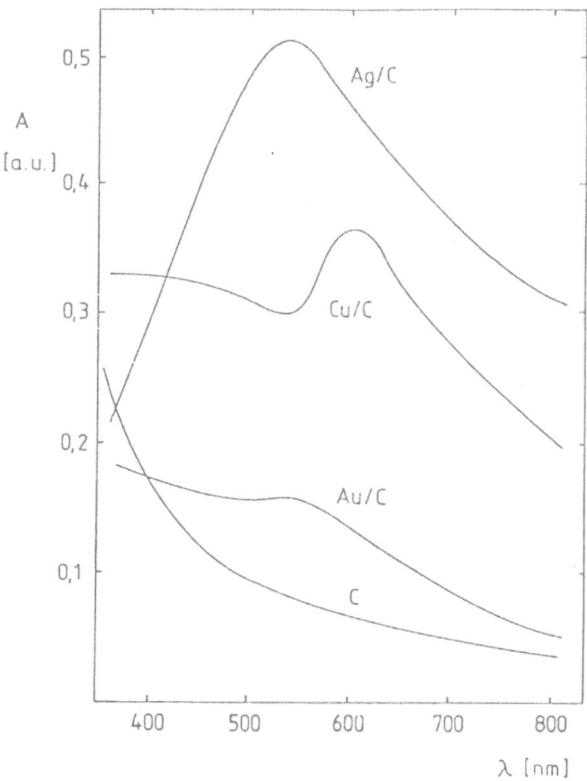

Fig. 4:

Optical absorbance in the visible region for a-C:H matrices (C) deposited at the same conditions as in Fig. 3 and containing different metals: silver (Ag/C), gold (Au/C) and copper (Cu/C).

A typical example showing how increasing concentration of metal affects the films' sheet resistance and microhardness is illustrated in Fig.3. The increasing amount of gold content is indicated by the ratio of emission line intensities, which correspond to the concentration of Au atoms and Ar in the gas phase, when a fixed butane/argon feed gas mixture is used. The sheet resistance subsequently exhibits a decreasing trend as the percolation threshold is approached. At the same time, the microhardness is also seen to decrease due to the introduction of deliberate "defects" in the form of metal clusters. The metastable a-C:H microstructure has been found to accommodate only a certain amount of the metal until its DLC structure is affected, following which it collapses, forming a predominantly graphitic matrix and hence softer deposits (5,18,34).

The presence of metal clusters in a-C:H matrices modifies the optical properties in the visible frequency region, due to anomalous absorption of the transmitted light which is "filtered". This results in different colorations of the composites as previously shown for the case of metal-filled plasma polymers (5). We have described in more detail elsewhere (5) how thee optical constants very sensitively depend upon the volume fraction, cluster size and shape of the metallic inclusions. The effect of different metals on optical absorption in the visible region is shown in Fig.4. Recently we have shown that these properties are stable with time, even at elevated temperature (35).

Conclusions

The investigation of a-C:H and metal-filled a_C:H films gas achieved significant progress with respect to film structures and properties, both from basic and applied points of view. Nevertheless, microscopic processes in plasma - surface interactions which lead to these remarkable carbonaceous deposits, constitutes an area of study in which much remains to be done.

Acknowledgement

The author wishes to express thanks to his former colleagues at Charles University, Prague: Prof. H. Biederman and drs. A. Fejfar, S. Néspurek, D. Slavínská and J. David for fruitful cooperation. The author is particularly grateful to Prof. M.R. Wertheimer (Ecole Polytechnique, Montreal) for encouragement, helpful discussions and for reading the manuscript.

References

1. J.C. Angus, P. Koidl and S.Domitz, in Plasma Deposited Thin Films, J. Mort and F. Jansen eds., CRC Boca Raton, 1986

2. P. Koidl and P. Oelhafen, eds., Proc. E-MRS Meeting, Vol.XVII, strasbourg, June 1987, Les editions de Physique, Paris, 1987

3. H. Biederman, L. Martinu, and J. Zemek, Vacuum 35 (1985) 447

4. H. Pagnia, J. colloid Polym. Sci., (1989), in press

5. H. Biederman, L. Martinu, in Plasma Deposition of Polymers (R. d'Agostino, ed.), Academic press, 1989 or H. Biederman, L. Martinu, D. Slavinska and I. Chudacek, Pure and Appl. Chem. 60 (1988) 607

6. H. Schmellenmeier, Z. Phys. Chem., 205 (1955/56) 349

7. H. Pagnia, Phys. Stat. Sol., 1 (1961) 499 and 2 (1962) 349

8. R.C. DeVries, Ann. Rev. Mater. Sci., 17 (1987) 161

9. C.P. Chang, D.L. Flamm, D.E. Ibbotson and J.A. Mucha, J. Appl. Phys, 63 (1988) 1744.

10. A.G. Whittaker, Science, 200 (1978) 763

11. R.B. Heimann, J. Kleiman and N.M. Salansky, Nature, 306 (1983) 164

12. L. Holland and S.M. Ojha, Thin Solid Films, 58 (1979) 107

13. A. Bubenzer, B. Dischler, G. Brand and P. Koidl, J. Appl. Phys., 54 (1983) 4590

14. Y. Catherine and P. Couderc, Thin Solid Films, 144 (1988) 265

15. B. Chapman, Glow Discharge Processes, Wiley Interscience, New York, 1980

16. K. Kohler, J.W. Coburn, D.E. Horne and E. Kay, J. Appl. Phys., 57 (1985) 59

17. J. Wagner, CH. Wild, F. Pohl and P. Koidl, Appl. Phys. Lett., 48 (1986) 106

18. H. Biederman, Hong Jon Chjok, L. Martinu, J. David, S. Kadlec, and P. Lukac, Pag.449 in ref.2

19. D.S. Whitmell and K. Williamson, Thin Solid Films, 35 (1976) 255

20. B. Window and N. Savvides, J. Vac. Sci. Technol. A, 4 (1986) 453

21. J. Kieser and M. Neusch, Thin Solid Films, 118 (1984) 203

22. T. Miyazawa, S. Misawa, S. Yoshida and S. Gonda, J. Appl. Phys., 55 (1984) 188

23. Ch. Weissmantel, Proc. IX Int. Vac. Congr. and V Int. Conf. Surf. Sci., Madrid, 1983, Pag.229

24. J. Franks, T.L. Ng and A.C. Wright, Vacuum, 38 (1988) 749

25. F.W. Smith, J. Appl. Phys., 55 (1984) 764

26. Ch. Weissmantel, Proc. ISIAT'83 and IPAT'83, Kyoto 1983

27. J.C. Angus, Thin Solid Films, 142 (1986)145

28. J.C. Angus, J. Vac. Sci. Technol. A, 6 (1988) 1778

29. J. Robertson and E. P. O'Reilly, Phys. Rev. B, 35 (1987) 2946

30. J. Robertson and E. P. O'Reilly, Pag.259 in ref. 2

31. D.I. Jones and A.D. Stewart, Phil: Mag. B, 46 (1982) 423

32. R.E. Sah, B. Dischler, A, Bubenzer and P. Koidl, Appl. Phys. Lett., 46 (1985) 739

33. R. d'Agostino, L. Martinu and V. Pische, Plasma Chem. Plasma Process., submitted.

34. Ch. Weissmantel, E. Ackerman, K. Bewilogua, G. Hecht, H. Kupfer and B. Rau, J. Vac. Sci. Technol. A, 4 (1986) 2982

35. H. Biederman, I. Chudacek, D. Slavinska, L. Martinu, J. David and S. Nespurek, Vacuum, in press

MATERIALS PROCESSING WITH PULSED ELECTRON BEAM

E. D'Anna, G. Leggieri, A. Luches and M. Martino

Dipartimento di Fisica, Università di Lecce
73100 Lecce, Italy

Abstract

Pulsed electron beam applications for materials processing are described. In particular, we review and discuss the relevant results obtained in our laboratory in the fields of synthesis of metal silicides, silicon carbide and alloying of immiscible metals.

Introduction

Energy pulses (produced by laser, electron and ion beams) are increasingly used for processing of materials, specially for removing lattice damages produced by ion implantation in semiconductors (1), for promoting metal-silicon reactions (2), metal-metal mixing (3,4) and synthesis of compound materials (5). Intense and short (a few tens of ns) energy pulses allow localized heating within a few microns below the surface, leaving unaffected the bulk of materials. Pulse heating can be also used to study the physical mechanism responsible for the reactions. Moreover, the fast heating and cooling rates (10^{10} K/s) allows the formation of highly metastable phases and the possibility of alloying immiscible metals. Pulsed electron and ion beams present some advantages with respect to the most commonly used laser beams, since:

a - The absorption of energy from particle beams is more
    predictable and less dependent on uncontrollable variables
    (surface characteristics, phase transformation, etc.);

b - it is possible to define the depth of the process from simple
    range-energy relations or, more accurately, from the spatial
    distribution of the energy loss, obtain through Monte-Carlo
    computations (6);

c - the overall efficiency is much higher.

Pulsed electron beams allow higher powers and higher energy densities than those allowed by pulsed ion beams. By using pulsed electron beams it is possible:

i. to process large areas (tens of cm²) in a single shot, with
    good uniformity of the process:

F. Garbassi and E. Occhiello (eds.), High Energy Density Technologies in Materials Science, 89–103.
© 1990 *Kluwer Academic Publishers.*

ii. to vary over relatively broad ranges the current, energy and pulse-width of the beam and the energy fluence on the samples.

Moreover pulsed electron beam machines are much cheaper to buy and easier to design and build than ion beam sources.
In contrast, electron beam sources ate more complex and sometimes more expensive than laser sources. Also, the maintenance may be more troublesome. Moreover, materials must be processed in vacuum.
In the following section we present a description of the electron beam generators, an outline of the methods most frequently used to determine the energy deposition of electrons in materials, the subsequent thermal effects in the irradiated samples and, finally, we present also a review of the most relevant results obtained in our laboratory in the field of processing of thin film with electron pulses.

Low energy electron beam system

Pulsed electron beams specially devoted to semiconductor water annealing were first developed at SPIRE Co. by A.R. Kirkpatric and co-workers (7). They claim that their patent for the use of pulsed electron beams to regrow the crystal lattice damage produced by ion implantation in silicon predates the use of lasers for similar applications. A typical SPIRE apparatus consists of a coaxial capacitor which is charged to a relatively high voltage (100-400 kV) with an electrostatic generator and then discharges into a field emission diode.
The electron beam accelerator used in our experiments has been described in previous papers (8-10). It consists (Fig.1) of a high voltage pulser, a coaxial Blumlein line and a field emission diode (FED). The high voltage pulser is a two stage Marx circuit, which supplies 10 to 50 kV pulses, depending on the charging voltage and spark-gap pressure.
In order to obtain a time compression of the voltage pulse, the Marx circuit is coupled to a Blumlein line, which gives the pulse an almost square shape and a length of about 50 ns (FWHM). This short pulse is applied to the cathode of a FED. The FED consist of a chatode and an anode contained in a vacuum vessel, evacuated at a pressure of 10-2 Pa, or better, by means of a rotary and an oil diffusion pumps. Graphite rods 10-50 mm in diameter are used as cathode. The cathode surface is finely machined to obtain good beam homogeneity. In fact, electron current emission occurs from whiskers protruding from the cathode surface. Early in the voltage pulse, these whiskers vaporize by Joule effect heating. If a dense array of whiskers is present, a uniform plasma sheet is provided and homogeneous beams are extracted. The anode of this diode consists of a highly transparent mesh, which allows the field-emitted electrons to propagate through and drift to the target to be irradiated.

Energy deposition in materials

The effect of pulsed electron beams irradiation on materials depend obviously on the energy fluence. These effects must be foreseen before submitting a sample to electron irradiation.

A quantitative description of the energy deposition process in materials requires good informations about the scattering, the energy loss and the range of the impinging electrons. These informations can be obtained either by means of theoretical models or by experiments. The results from theoretical models are available as more or less complicated analytical formulas (11) and the models are based on many simplifying assumptions (like, for example, the target homogeneity). Thus, they can be used only in special cases. When the target is not uniform along the direction of propagation of the beam, it was known that statistical models give better informations about the electron penetration in solid materials. The statistical method which was shown to give very good results is based on the Monte-Carlo (MC) techniques (6). The model used in our works for MC simulation of electron scattering and energy deposition into the target has been described in detail elsewhere (12). Here only a brief summary is presented.

The sequence of events for one electron with energy E_o impinging on a target is as follows. The first scattering event is assumed to occur at the surface of the irradiated sample. The electron path is determined by successive single scattering into the material. For each scattering the effective path length S of the electron, the scattering angle θ and the azimuthal angle Φ are derived from the differential cross section for elastic scattering, as given by the screened Rutherford formula. Then, the path length S is calculated by equating the probability for an electron to travel over a distance S between two collisions to a uniform pseudo-random number 0<R≤1. The angles (θ and Φ) are chosen by equating two other uniformly distributed pseudorandom numbers to the probability for an electron to be scattered into a definite solid angle with azimuthal symmetry. If the target is a compound, another pseudorandom number is used to determine which atomic species acts as the scattering centre.

The mean energy loss between two successive elastic scattering events is calculated by using the Continuous-Slowing-Down- Approximation (CSDA) (13). Thus, the energy loss in an elemental volume in the target is given by the well known Bethe formula (taking into account the Rao-Sahib and Wittry modifications (14)). Obviously, in the case of layered materials, all the chemical-physical parameters and the mean free path are varied when the electron moves across the interfaces. The calculation is repeated to yield new positions for a new set of pseudorandom numbers until the electron energy falls below a predesignated vale (E_{min}).

To simulate the effect of a real electron beam, a large number of incident electrons must be utilized in calculations (1000 in the following computations). As an example, the energy loss diagrams obtained with our MC code for crystalline silicon and for the layered system Pt(300 nm)/Si are represented in Figures 2 and 3.

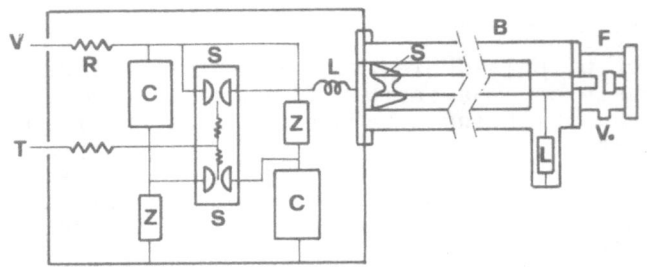

Fig. 1. Outline of our electron beam generator. V: input voltage;
T: trigger; R: charging resistors; C: capacitor; Z: imped-
ence; S: spark-gap; L: inductor; B: Blumlein line; F: field
emission diode; V.: vacuum port.

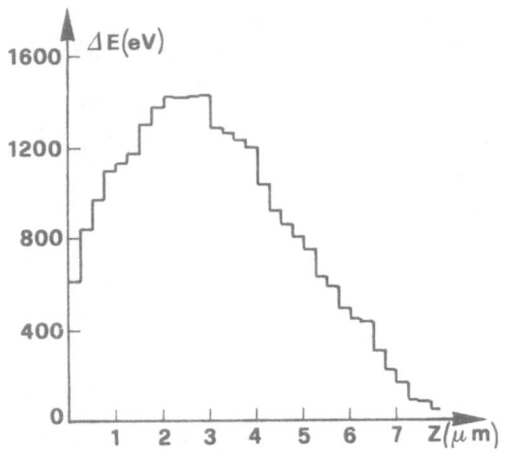

Fig. 2. Diagram of the energy loss per layer (ΔE) of 30 keV electrons impinging on a crystalline silicon sample.

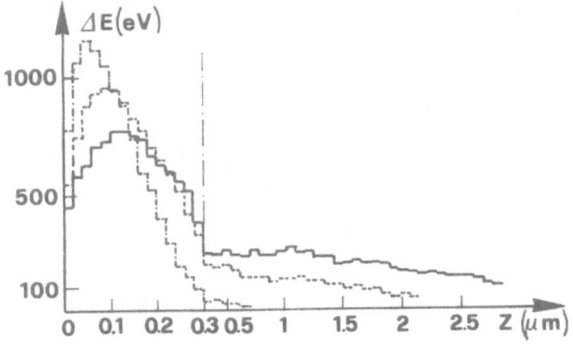

Fig. 3. Diagram of the energy loss per layer (ΔE) of 18 keV (-·-),
25 keV (---) and 30 keV (——) electrons.

Temperature profiles into materials

To increase our understanding of the thermal effects produced by an incident pulsed electron beam on materials, it is important to control not only the energy fluence, but also the temperature of the irradiated layers, the temperature of the interfaces and the time evolution of the temperatures.

The thermal effects produced by electron pulse can be determined by solving the heat diffusion equation (15). Assuming that a) the electron density is uniform during the pulse time, b) the section of the electron beam interacting with the solid surface is greater than the penetration depth of heat on the time scale of interest, and c) the composition of the sample is homogeneous in all directions except in the direction of penetration of the beam, the heat diffusion equation can be written as follows:

(1) $\quad \varrho(z,T)c(z,t) \, \delta T/\delta t = \delta z[k(z,T)\delta T/\delta z] + W(z,t)$

Where z is the direction of propagation of the beam; $\varrho(z,T)$, $c(z,T)$ and $k(z,t)$ are the density, the specific heat capacity and the thermal conductivity, respectively, at position z and temperature T; $W(z,t)$ is the energy loss rate per unit volume, determined through MC computations.

$W(z,t)$ can be written as

(2) $\quad W(z,t) = W(z)F(t)$

where $F(t)$ is the pulse shape function.
$F(t) = A \exp[-\alpha(t-t_o)^2$ is the function that, with a proper choice of the constants, approximates very well the actual shape of our electron pulses.

Equation (1) can be solved by means of numerical methods (16), taking account the phase transitions and the following boundary conditions:

$(T/z)_{z-o} = 0$ (no heat dissipation from the surface)
$T(z,0) = To$ for all z
$T(L,t) = To$ (good thermal contact with a thermostat)
$(k \, \delta T/\delta z)_{z_i} = (k \, \delta T/\delta z)_{z_i}$ (good thermal contact at interface)

L is the sample thickness and z_i is the interface position.

As an example, in Figs. 4 and 5 the temperature profiles and the interface temperature evolution in the case of a Pt(300 nm)/Si sample submitted to a single 25 keV - 50 ns electron pulse at various current densities are shown.

In our computer code it is possible to take into account also the heat of formation of the compounds synthesized as a consequence of the energy pulse. The result for a Pt(300 nm)/Si sample submitted to a 25 keV - 50 ns - 500 A/cm² pulse, is shown in Fig. 6. By comparing Fig.6 with the curves of Fig. 5 corresponding to the same current density, it is evident that the thermal effects are due mainly to the energy deposited by the beam.

Application

Our pulsed electron beam was used to obtain:

1 - metal/silicon (M/Si) reactions;
2 - carbon/silicon (C/Si) reactions, and
3 - metal/metal mixing.

M/Si reactions

The impetus to study M/Si reactions is due to the require-ments placed on the performance of integrated circuits, where silicides are used to produce ohmic and rectifying contacts, barriers to interdiffusion, stable and reactive overlayers, interconnects and so on (17).

We reacted near-noble and transition metals, deposited mainly on Si single crystals with thickness of the M layer of 300 to 300 nm. Near-noble and transition metal silicides are of great interest in integrated circuit technology.

The effects of the irradiations were studied by 2 MeV He$^+$ Rutherford backscattering spectrometry (RBS) measurements. X-ray and electron diffraction techniques were used to positively

Tab. 1 : Summary of the Metal/Silicon systems irradiated with electron beam pulses.

System	Observed compounds	Energy density range (J/cm^2)
Pt/Si	$PtSi$, Pt_2Si, Pt_3Si	0.4 - 1.8
Pd/Si	$PdSi$, Pd_2Si, Pd_3Si, Pd_4Si	0.4 - 1.9
Ni/Si	$NiSi_2$, $NiSi$, Ni_2Si	1.5 - 1.8
Cr/Si	$CrSi_2$	1.2 - 1.5
Mo/Si	$MoSi_2$, Mo_5Si_3	2.4 - 3.4
W/Si	WSi_2, W_5Si_3	2.5 - 3.5
Nb/Si	$NbSi_2$, Nb_5Si_3	2.4 - 3.2
Ti/Si	$TiSi_2$, $TiSi$, Ti_5Si_3	2.6 - 3.9
V/Si	V_6Si_5, V_3Si	2.2 - 3.1

identify the crystallographic phases formed in the intermixed layers and scanning electron microscopy (SEM) was used to investigate surface structures. The results obtained in our experiments are summarized in Table 1.

As a general feature, M_2Si and MSi_2 are the dominant phases for near-noble and refractory metals, respectively, formed at relatively low electron fluences. At high fluences the irradiated layers are not homogeneous in depth and most of the phase predicted by the phase diagram are simultaneously detected. The compounds richest in Si seem to be located in the innermost side of the reacted layer and the ones richest in metal in the outermost side. Thermal post-annealing of irradiated samples in conventional vacuum furnaces proved the thermal instability of many of the phases formed during the energy pulses (18).

Generally, the high energy delivered by the electron beam pulse raises the temperature in the film to values high enough for liquefaction to occur. The high diffusivity in the liquid produces the intermixing between silicon and metal. The subsequent cooling down is responsible for the phase formation. However, sometimes reactions were observed at very low fluences, certainly not sufficient to melt the metal film. It was also observed that a correlation exist between the minimum energy density necessary to produce a measurable reaction and the lowest eutectic temperature T_e of the binary phase diagram for all the investigated M/Si system (19). For example, in the Pt(300 nm/Si samples, we observed that the RBS spectra show evidence of reaction at as low an electron current density as 400 A/cm², corresponding to an energy density of 0.5 J/cm² (Fig.7).

At this low fluences, the thin reacted layer shown a composition close to the lowest eutectic (77% Pt, 23% Si) of the binary Pt-Si system. After a 500 A/cm² (0.6 J/cm²) pulse, an almost uniform layer of the Pt_2Si silicide is formed at the Pt/Si interface. After a 750 A/cm² (0.9 J/cm²)pulse, all the Pt film has reacted with Si to form a mixture of silicedes (PtSi, Pt_2Si, Pt_3Si). After a 1000 A/cm² (1.25 J/cm²) pulse, non uniform diffusion of Pt in SI and surface damage is observed.

From the calculated temperature profiles and evolutions (Figs. 4 and 5), we can observe that no melting of the components should occur up to a current density of about 700 A/cm², when Si starts melting. The Pt surface exceeds the Pt melting temperature at about 750 A/cm², and it melts down to the interface when the current density exceeds 800 A/cm².

A similar behaviour was observed also in Cr/Si samples. AS an example, from the RBS spectra (Fig.8) of Cr(65 nm)/Si samples, we can observe that no reaction is produced at a current density if 950 A/cm² (1.2 J/cm²). At 1000 A/cm² a $CrSi_2$ layer about 70 nm thick is formed, while about 50 nm of Cr remain unreacted at the surface. The reacted layer grows up as the current density increases, until at 1200 A/cm² (1.5 J/cm²) the whole Cr film has reacted.

If one correlates the experimental results to the temperature profiles and evolutions obtained by solving Eq. (1), it results that silicide formation starts at temperature lower than the melting temperature of Si and that melting temperature

96

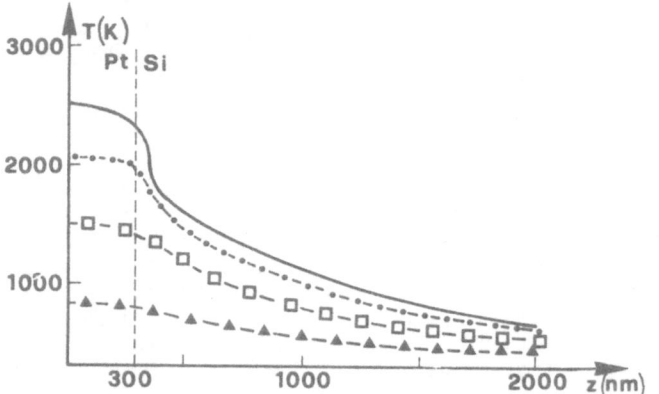

Fig. 4. Temperature profiles in the Pt(300 nm)/Si sample as a func-
tion of depth. $-\blacktriangle-$ 250 A/cm^2; $-\square-$ 500 A/cm^2; $-\cdot-$ 750 A/cm^2;
$\longrightarrow$ 1000 A/cm^2.

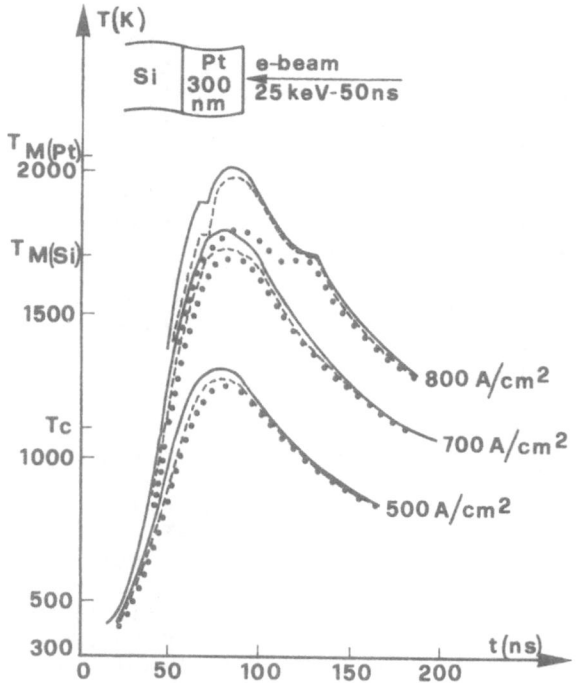

Fig. 5. Time evolution of the temperature in a Pt(300 nm)/Si
sample, submitted to a 25 keV -50 ns pulse. $\longrightarrow$ z = 30 nm;
$---$ z = 270 nm; $\bullet\bullet\bullet$ z = 330 nm.

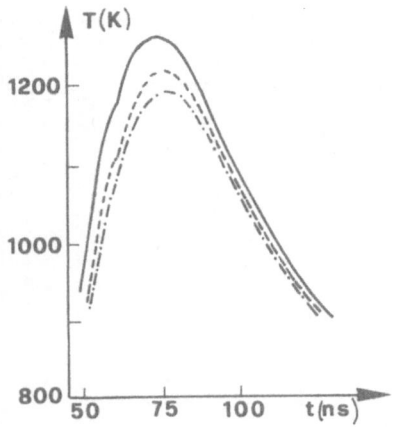

Fig. 6. Time evolution of the temperature of a Pt(300 nm)/Si sample irradiated at 500 A/cm² (the free energy of formation of Pt₂Si is taken into account). —— z = 30 nm; --- z = 270 nm; -·- z = 330 nm.

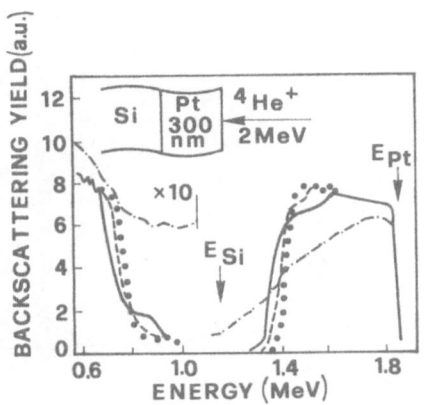

Fig. 7. RBS (2 MeV, He⁺) spectra of Pt(300 nm)/Si samples. (···) as deposited; after irradiation (25 keV, 50 ns) at (---) 400 A/cm²; (——) 500 A/cm²; (-·-) 750 A/cm²

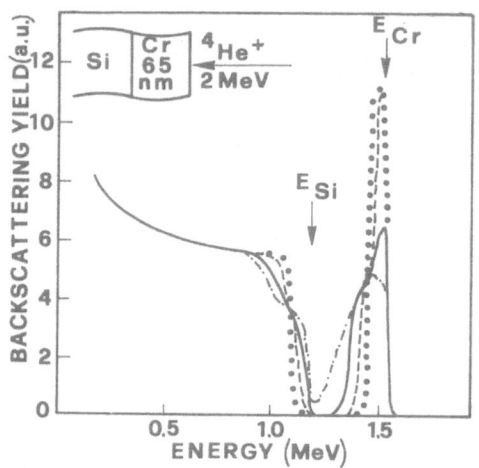

Fig. 8. RBS (2 MeV,He⁺) spectra of Si/Cr(65 nm) samples. (···) as deposited; after irradiation (25 keV, 50 ns) at (---) 1000 A/cm²; (—) 1100 A/cm²; (-·-) 1200 A/cm².

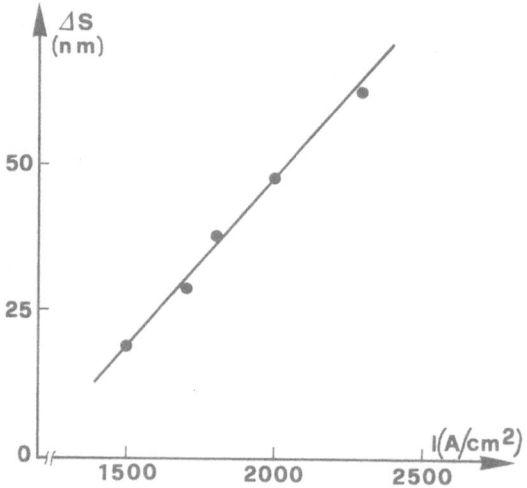

Fig. 9. Thickness ΔS of the Si-C mixed layer vs electron beam density (25 keV - 50 ns). The dots represent experimental results. The solid line represents a fit, to show the linear (within the experimental errors) dependence of ΔS with the electron current density.

of Cr should be reached not even at the highest fluence used in
this experiment (1200 A/cm²). The fact that melting of the thin
Cr film has not occurred is confirmed by SEM micrographs of the
irradiated samples.

Even at the highest fluence the surface does not exhibit the
characteristic ripples that appear when a melt is rapidly cooled
down. Only some cracks are present where the high thermal and
mechanical stress, due to the melted silicon underlayer, caused
local breaking of the thinned chromium film.

We found a similar behaviour also in the Mo/Si, Ti/Si and
W/Si systems. Details on the experimental results and calculated
temperature profiles and evolutions are extensively described in
topic works (20-25).

As a conclusion, from our experimental results, compared to
the temperature computation, we can deduce than silicide
formation starts at a temperature lower than the melting
temperature of the metal film; the reactions start when the
interface temperature exceeds the lowest eutectic temperature of
the M/Si system and the growth rate of the silicides is typical
of a liquid phase (1 m/s) even if melting of the elements is not
reached.

C/Si reactions

Silicon carbide (SiC) was also aynthesized using electron
beam pulses. SiC is at present the subject of intense studies for
its potential applications to many fields of high technoligical
interest as surface coatings with high wear resistance and as
adhesion promoters (26). Microwave devices, blue light emitting
diodes and sensor capable of high temperature operation are also
expected from this kind of ceramic material (27). Several
deposition techniques are currently used to produce SiC films,
like CVD, reactive evaporation and RF cathode sputtering (28). To
form SiC layers on Si, heteroepitaxial growth of SiC on Si
substrates at high temperature, ion implantation og hogh dose of
C to Si substrates and ion mixing of C film deposited on Si
underlayers, followed by thermal annealing, are possible
techniques (29). However, all these procedures present
experimental difficulties. Moreover, the very high tempreature
required to obtain crystalline SiC films can cause a number of
drawbacks, like impurity diffusion.

Due to the above difficulties, we tried a different approach
to study the synthesis and crystalline evolution of SiC films.
Pulsed electron beams were used to react thin C films deposited
on Si single crystal wafers. The short pulses allow localized
heating of the samples, thus avoiding impurity diffusion.

Thin films of C (100 to 500 nm thick) were sputtered in pure
Ar gas on single crystal <100> wafers. The samples were
irradiated with 25 keV - 50 ns electron beam pulses at various
current densities.

RBS analysis shows that in C(100 nm)/Si samples mixing
starts at current densities higher han about 1200 A/cm² (1.5
J/cm²). The thickness of the intermixed layer increases almost

linearly with the current densities (Fig. 9). At current higher than 2400 A/cm² ablation of the C film is observed. Thicker C layers (400 nm and 500 nm) were also reacted with the Si substrate. The threshold current density for starting the reaction is almost independent of the C film thickness. From The RBS spectra a composition of the mixed layer close to the SiC compound was deduced. TEM and electron diffraction analysis of irradiated samples shows that the composition of the mixed layer is hexagonal α-SiC with lattica parameters a = 0.3703 nm and c= 1.00053 nm (30).

M/M Mixing

An alternative approach to ion implantation for the formation of a surface alloy of a given composition between immiscible metals is to deposit a thin metal layer on a proper substrate and then to induce an atomic scale mixing between the film and the substrate.

We performed a study on the Pb-Al system (31) and on the Ni-Ag system (32), irradiated with high current electron beam pulses. The Pb-Al equilibrium phase diagram shows practically no solid solubility of Pb in Al and no intermediate phase. The effects of pulsed electron beam irraiation (pulse width 50 ns FWHM, electron energies from 15 to 30 keV and current densities in the range 100-2300 A/cm²) on Al-Pb systems prepared by vacuum evaporation of Pb layers (50 to 200 nm) over Al single-crystal underlayers, have been investigated by RBS and SEM techniques. We obtained surface alloys with up to 3 at.% Pb in Al, when the current density was equal to 2300 A/cm².

The second system, Ni-Ag, was irradiated with a single electron beam pulse (pulse width 50 nm, electron energy 25 keV and peak current densities in the range 600 to 2500 A/cm²). The samples were prepared by vacuum evaporating 70 nm and 185 nm thick Ag layers on Ni single crystals. The thermal equilibrium phase diagram of the Ag-Ni system shows a wide miscibility gap both in solid and liquid phase and no intermediate phase. The maximum solid solubility ia about 0.1% Ni atoms at the Ag-rich side (at 960 C) and about 2% Ag atoms at the Ni rich side (at 1435 C). The RBS spectra and SEM pictures of the irradiated samples indicated that extended solid solubilities were achieved. Maximum solubility was determined to be 16 at.% Ag in Ni and 4.5 at.% Ni in Ag. At the higher current densities mixing is accompanied by a strong Ag loss in the 70 nm Ag case (up to 50% of the deposited layer).

Conclusions

Pulsed electron beams are a useful tool for heating and melting of surface layers. By using this technique, it is possible to obtain:

a) regrowth and electrical activation of ion implanted

semiconductors,

b)   formation of metal silicides,

c)   formation of metastable compounds,.

d)   formation of compound materials,

e)   alloying of immiscible metals.

It must be noted that thremally metastable phases cannot be produced by conventional thermal annealing. Also refractory metal silicides of great interest in integrated circuit technology (like $TiSi_2$ and WSi_2) are easily formed by pulsed electron beam annealing. The above mentioned silicides, due to their high formation temperature, cannot be prepared by furnace annealing of M/Si systems. They are usually formed through co-sputtering techniques, followed by high temperature heating in a vacuum furnace. The easiness of control and versatility of beam parameters make pulsed electron beams a unique tool for materials processing.
The main problem in materials processing with pulsed electron beams is the beam spatial uniformity, which can be obtained through a careful control of the cathode of the electron source (19).

Acknowledgements

Work sustained in part by Ministero della Pubblica Istruzione. We thank F. De Donno and L. Villa for the drawings. The analysis of the irradiated samples were carried out through collaborations with researchers of the Universities of Ancona, Modena and Padua.

References

1.   D. K. Biegelsen, G. A. Rozgonyi and C. V. Shank, Eds., "Energy Beam Solid Interactions and Transient Thermal Processing", Mater. REs. Soc. Symp. Proc. Vol. 35 (1984)

2.   C. W. White and P. S: Peercy, Eds., "Laser and Electron Beam Processing of Materials", Academic Press, New York, 1981

3.   J.M. Poate, G. Foti and D.C. Jacobson, eds., "Surface Modification and Alloying", Plenum Press, New York, 1983

4.   J.F. Gibbons, L.D. Hess and T.W. Sigmon, eds., "Laser and Electron beam Solid Interaction and Material Processing", North Holland, New York - Oxford, 1981

5. V.T. Nguyen and A.G. Cullis, eds., "Energy Beam-Solid Interactions and Transient Thermal Processing", Les Editions de Physique, Strasbourg, 1985

6. D. Kyser and K. Murata, IBM J. Res. Dev. 18 (1974) 352

7. A.R. Kirkpatrick, J.A. Minnucci and A.C. Greenwald, I.E.E.E. Trans. Electron Devices ED-24 (1977) 429

8. A. Luches, V. Nassisi, A. Perrone e M.R. Perrone, Physica 104C (1981) 228

9. G. Leggieri, A. Luches, V. Nassisi, A. Perrone, M.R. Perrone, G. Majni and F. Nava, Vacuum 32 (1982) 9

10. E. D'Anna, G. Leggieri, A. Luches, V. Nassisi, A. Perrone, M.R. Perrone and R. Simmini, Vacuum 35 (1985) 19

11. K. Knaya and S. Okayama, J. Phys. D: Appl. Phys. 5 (1972) 43

12. E. D'Anna G. Leggieri, A. Luches, M. Martino and V. Nassisi, Vacuum 38 (1983) 175

13. R.D. Birkhoff, "Handbook der Physik 34". Springer Verlag, Berlin 1958

14. T.S. Rao-Sahib and D.B. Wittry, in G. Shinoda, K. Kohra and T.Ichinokawa, eds., Proc. 6th Int. Conf. on X-ray Optics and Microanalysis, University Tokyo Press, Tokyo 1972

15. H.S. Carlus and J.C. Jaeger, "Conduction of Heat in Solids", Clarendom Press, Oxford, 1959

16. D.U: von Rosemberg, "Methods for Numerical Solutions of Partial Differential Equations, American Elsevier Publishing Co. INc., New York 1969

17. K.N. Tu and J.W. Mayer, in J.M. Poate, K.N. Tu and J.W. Mayer eds., "Thin films Interdiffusion and Reactions", Wiley, New York 1978

18. G. Majni, F. Nava, G. Ottaviani, E. D'Anna, G. Leggieri, A. Luches and G. Celotti, J. Appl. Phys. 52 (1981) 4055

19. G. Ottaviani. J. Vac. Sci. Technol. 16 (1979) 1112

20. E. D'Anna, G. Leggieri, A. Luches, G. Majni, G. Ottaviani and M.R. Perrone in V.T. Nguyen and A.G. Cullis, eds., "Energy Beam-Solid Interactions and Transient Thermal Processing", Les Editions de Physique, Paris, 1985

21. E. D'Anna, G. Leggieri, A. Luches, G. Majni and G. Ottaviani, Thin Solid Films 136 (1986)163

22. E. D'Anna, G. Leggieri, A. Luches, G. Majni, F. nava and Ottaviani, Appl. Phys. A40 (1986) 183

23. E. D'Anna, G. Leggieri, A. Luches and G. Majni, Thin Solid Films 140 (1986) 163

24. E. D'Anna, G. Leggieri, A. Luches and G. Majni, J. Vac. Sci. Technol. A5 (1987) 1726

25. E. D'Anna, G. Leggieri, A. Luches and G. Majni, Le vide-Les Couches Minches, 42 (1987)67

26. S. Inoue, K. Yoshii, M. Umeno and H. Kawabe, Thin Solid Films 151 (1987) 403

27. I. Ohdomari, S. Sha, H. Aochi, T. Chickyow and S. Suzuki, J. Appl. Phys. 62 (1987) 3747

28. W. Weisweier, G. Nagel and J. Klepp, Thin Solid FIlms 155 (1987) 39

29. T. Kimura, H. Yamaguchi, L. Ji-Kui, S. Yugo, Y. Adachi and Y. Kazumata, Thin Solid Films 157 (1987) 117

30. E. D'Anna, G. Leggieri, A. Luches, V. nassisi, A. Perrone, G. Majni and P. Mengucci, Applied Surface Physics (in press)

31. G. Battaglin, A. Carnera, L.F. Donà dalle Rose, V. N. Kulkarni, P. Mazzoldi, E. D'Anna, G. Leggieri and A. Luches, Mat. Res. Soc. Symp. Proc., 23 (1984)769

32. G. Battaglin, A. Carnera, L.F. Donà dalle Rose, P. Mazzoldi, E. D'Anna, G. Leggieri and A. Luches, Thin Solid Films 145 (1986) 147

ION BEAM TREATMENT OF POLYMERS

E. Occhiello, F. Garbassi

Istituto Guido Donegani S.p.A.
Via G. Fauser 4
28100 Novara ITALY

Abstract

An overview of the application of ion beam treatments to polymers is presented. Low energy ion irradiation has been used to improve the efficiency of dry etching processes.,and adhesion. High energy irradiation (> 10 keV) results in drastic changes in chemistry and structure of treated polymers. Changes in solubility, conductivity and refractive index have been observed.

Introduction

Ion beam treatments have rapidly become a subject of high academic and industrial interest (1-3). Ion beam processes, depending on their energy, fall into two main categories. The first includes low energy ions (up to 5-10 keV), normally used in ion beam etching and ion-assisted deposition. The second is related to high energy ions (more than 10 keV, up in the MeV region) used mainly for ion implantation processes.

In the case of polymers this same distinction holds, the scope and application of low and high energy ions is very different. A fundamental consideration is penetration, as shown in Tab. 1 (4).

A first section in the following is devoted to low energy ions. In this case the penetration is low (a few nanometers), therefore the effect is on very thin surface effect. The affected properties are etch rate in dry etching environments and adhesion.

A second section is related to the effect of high energy ion irradiation. In this case the effect is on a layer which may be up to some micrometers thick. Therefore effects on solubility of thin films, conductivity and light transport are present (1-3).

Low energy ions

The most important application in which ion bombardment of polymers is involved is in dry etching for microelectronic applications. Resist removal can be performed in RIE (Reactive Ion Etching) or even in RIBE (Reactive Ion Beam Etching) conditions. A number of works have appeared in this area (5) and reference can be made within this book to the papers by d'Agostino et al. and Occhiello et al..

F. Garbassi and E. Occhiello (eds.), High Energy Density Technologies in Materials Science, 105–113.

Tab. 1

Total Ar ion ranges for polyethylene (PE), polymethylmethacrylate (PMMA), polyimide (PI) and polytetrafluoroethylene (PTFE) as a function of ion energy.

Total Ar ion ranges (nm)

Energy (keV)	PE	PMMA	PI	PTFE
0.1	1.79	1.66	1.39	1.28
0.5	4.04	3.72	3.15	2.84
1.0	5.89	5.42	4.60	4.12
2.0	8.80	8.10	6.88	6.14
5.0	15.68	14.42	12.32	10.93
10.0	25.32	23.31	19.98	17.70
30.0	59.03	54.59	47.08	41.82
50.0	90.92	84.40	73.01	65.17
70.0	122.35	113.92	98.76	88.58
100.0	169.07	158.01	137.31	123.97
300.0	467.14	442.65	387.59	359.72
400.0	606.35	576.54	506.00	473.04
500.0	739.39	704.75	619.70	582.27
800.0	1105.47	1056.95	931.82	885.16
1000.0	1323.20	1264.86	1115.84	1064.77

Fig. 1. Effect of 2 MeV Ar^+ irradiation of polymers vs. dose.

Increasing dose (ions/cm²)

10^{10} 10^{11} 10^{12} 10^{13} 10^{14} 10^{15} 10^{16} 10^{17} 10^{18}

<---------------------->
CHEMICAL ALTERATION, CROSS-LINKING

    <-------------------->
    CHANGES IN OPTICAL PROPERTIES

      <-------------------->
      CARBONIZATION

      <-------------------------->
      CHANGE IN CONDUCTIVITY

People involved in dry etching normally study etch rates as a function of etching environment and ion bombardment parameters (energy and current). Some studies have anyway appeared also on the effect of ion bombardment on surface chemistry (6-7), showing disruption of the original surface chemistry, with destruction of aromatic rings, introduction of oxygen as a consequence of reaction with atmosphere, removal of polar elements form the polymer, etc..

An interesting study about the effect of ion beam treatment on the surface crosslinking of polystyrene (PS) was performed by a group from Cornell Univ. (8). Ion beam treatment of a two-layered sample made of deuterated and non-deuterated PS allowed to observe the formation of a tightly crosslinked surface layer, which prohibited diffusion with the underlying layers.

Surface crosslinking affects surface mechanical properties and, along with surface chemistry and morphology, it determines adhesion. Actually it was shown that Ar ion beam etching improved rather effectively the adhesion of sputter deposited copper to polyimide (9).

An important improvement in adhesion was also observed as a consequence of Ar ion bombardment of HDPE (10). The effects of ion bombardment were dehydrogenation and crosslinking of the surface region, reaching the limit of graphitization at high ion dose. Important improvements in the adhesion of evaporated Ti films were observed after the treatment, the fracture energy reached the cohesive strength of HDPE itself.

High energy Ions

High energy ion beam treatment has been used mostly to change in a controlled way the chemistry of thin layers, thereby imparting desirable electrical, optical, tribological properties (1-3). The impact of highly energetic ions is known to induce a large amount of damage in the crystal lattice, thus requiring post-annealing steps to remove it.

Polymers and in general organic solids are more difficult to deal with. Their inherent instability to radiation and to high temperatures means that any ion beam treatment will anyway change the chemistry of the treated layer and that annealing cannot be used as efficiently as in metals and semiconductors to partially mend damages (11).

Of course the extent of damage is very much dose-related, and typical ranges for different phenomena occurring in polymer layer irradiated with 2 MeV Ar ions are presented in Fig. 1 (11). Chemical changes occur even at low doses, later on carbonization, changes in optical and electrical properties are induced.

The type of chemical changes of polymer substrates undergoing ion irradiation are summarized in Fig. 2. First of all a chemical change occurs in polymers containing heteroatoms (oxygen, nitrogen, fluorine, chlorine). Molecular species are emitted by all polymers as a consequence of irradiation and a

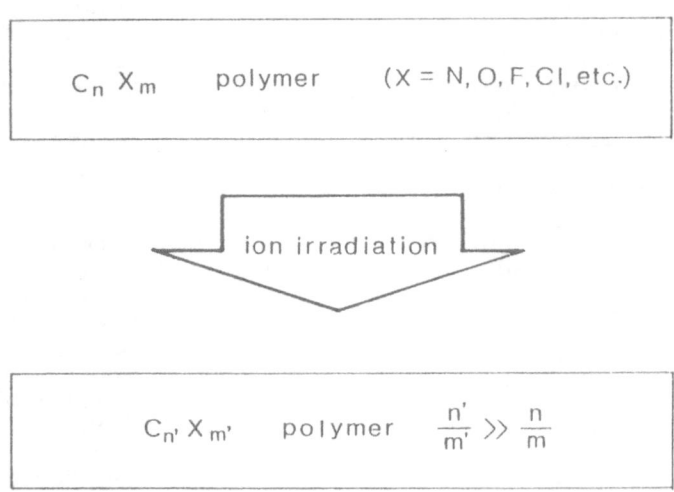

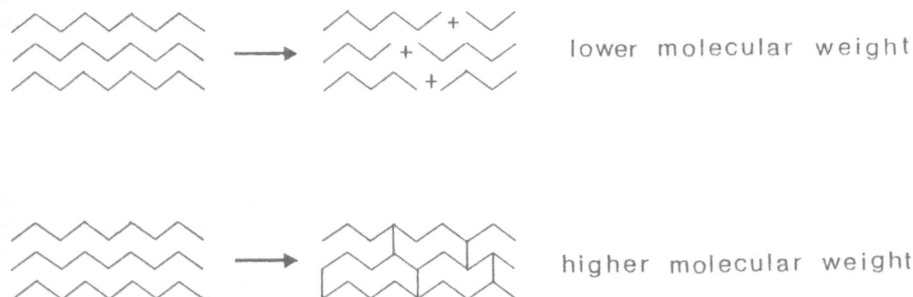

Fig. 2. Chemical modifications by ion beam irradiation:
a) stoichiometry change, b) scission and crosslinking.

general rule is that a depletion in heteroatoms occurs (Fig. 2-a), due to the formation of stable neutral molecules such as carbon oxides, HCN, HCl, HF, etc. (12).

Another typical effect of irradiation is a decrease in H/C ratio, due to efficient removal of hydrogen, leading, at very high ion dose, to graphitization of the irradiated layer (13).

All irradiations, using photons, electrons or ions, induce scission or crosslinking of polymer chains (Fig. 2-b). The molecular weight of the sample is changed depending on the chemistry of the polymer. As with electron irradiation negative resists, such as PS, are crosslinked. Positive resists, such as PMMA, are efficiently lowered in molecular weight, the dependence of the inverse molecular weight ($1/M_n^*$) upon the irradiation dose (Φ) is given below (14).

$$1/M_n^* = 1/M_n^\circ + 0.01 \; G \; \Phi \; N_A^{-1}$$

$1/M_n^\circ$ is the initial inverse molecular weight, G is the scission yield, defined as the number of bonds broken per 100 eV absorbed energy, and N_A is Avogadro's number.

Considerations about scission and crosslinking lead immediately to the most important possible application of ion beam irradiation to polymers, i.e. lithography. The comparison with electron beam lithography is particularly interesting. Ions have two advantages: resolution and sensitivity.

In electron-beam lithography the phenomena of lateral scattering and creation of energetic secondary electrons limit resolution and induce losses in contrast. The scattering of ions is much lower and only low energy secondary electrons are produced, thereby improving resolution (15). Furthermore the sensitivity of resists to electron beam irradiation is very much changeable (16), while the amount of ion energy absorbed per unit volume is relatively similar for a large number of polymers (17) allowing to choose resists on the basis of characteristics other than sensitivity.

Ion implantation has also been shown to modify drastically the electrical conduction properties of polymers. Irradiation with 10^{14}-10^{15} ions/cm^2 fluences (Fig. 1) has been shown to increase drastically the conductivity of polymers. In Fig. 3 the resistivity ranges of metals, semiconductors, insulators and 2 MeV Ar ion irradiated polymers (18) are shown.

Of course increases in conductivity are dependent on irradiation parameters, in particular fluence and ion energy. Everybody agrees that for many polymers conductivity increases nearly linearly with ion dose, to saturate at high fluences, typically above 10^{16} ions/cm^2 (19-21). Energy-wise a dependence of saturation resistivity on energy was observed. In particular MeV irradiated samples were shown to have lower resistivity values than 25-400 keV irradiated ones (18, 22). A rather extensive review of all studies related to conductivity enhancement by ion irradiation is presented in (5).

The change in optical properties of irradiated polymers have also attracted a number of studies (5). In general at low and intermediate fluences the results of ions irradiation are

110

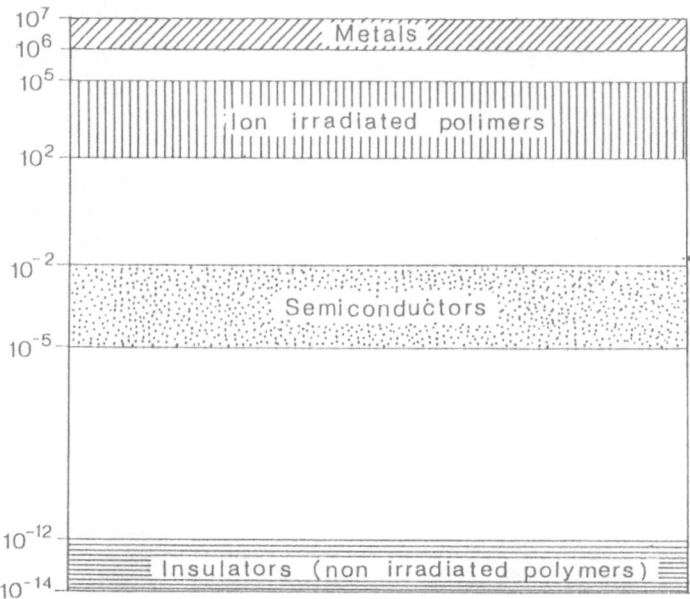

Fig. 3.   Conductivities (Ohm⁻¹cm⁻¹) of different materials.

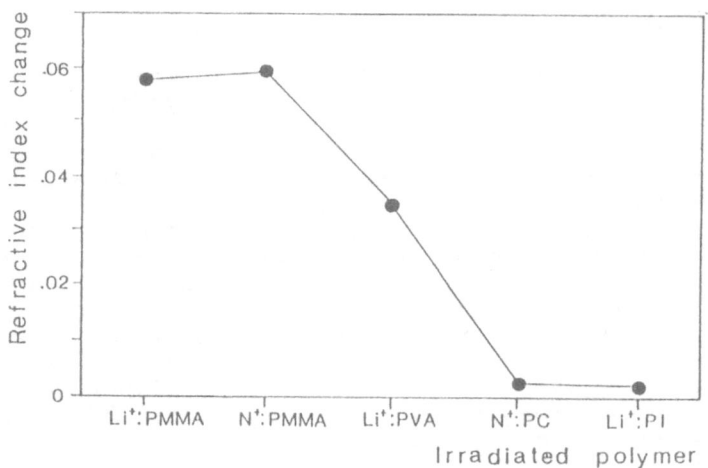

Fig. 4.   Refractive index change upon ion irradiation.

continuous colour changes. At higher fluences carbonization occurs and therefore the films turn opaque. A logical consequence is that ion irradiation can be used to alter refractive indexes.

Fig. 4 shows the change in refractive index of polymers irradiated with 100-130 keV Li^+ and N^+ ions (23). Implantation in aliphatic materials (PMMA, PVA) causes larger index changes than in aromatic polymers (PC, PI). A possible application consequent to the possibility of altering refraction indexes by irradiation is the direct recording of light guide patterns in polymers (23).

Conclusions

Ion beam irradiation of metal and semiconductor has already found sound commercial applications. In the case of polymers it is not so, part of the reason is that the control of the effect of the treatment is rather difficult. Ion irradiation tends to change drastically the chemistry and structure of polymers and no annealing procedure such as those performed on semiconductors and metals is possible.

Possible future applications are in the field of lithography, due to the possibility of finely focussing ion beams (resolution > 10 nm) and achieving advantages in resolution and sensitivity of resist materials. The possibility of playing with the transport of light might open up the field of storage of information as well.

Apart from electronics, some prospects are present in the field of biomaterials, for changing the properties of the surface layer to make them more compatible with biological fluids (24-25). Interestingly, even if ion beam treatments are normally found in high added value applications, a study on the treatment of fibres for composite materials by ion beam irradiation has been presented recently (26).

References

1. O. Auciello, R. Kelly, Eds., "Ion Bombardment of Surfaces", Elsevier, Amsterdam, 1985

2. P. Mazzoldi, G. W. Arnold, Eds., "Ion Beam Modification of Insulators", Elsevier, Amsterdam, 1987

3. P. Siohansi, Thin Solid Films, 118, 61 (1984)

4. J. P. Biersack, ref. 2, p. 648

5. J. W. Coburn, Plasma Chem. Plasma Proc., 2, 1 (1982) and refs. therein

6. W. E. Vanderlinde, P. J. Mills, E. J. Kramer, A. L. Ruoff, J. Vac. Sci. Technol., b4, 1362 (1985)

7. P. Bodo, J.-E. Sundgren, Thin Solid Films, **136**, 1147 (1986)

8. S. F. Tead, W. E. Vanderlinde, A. L. Ruoff, E. J. Kramer, Appl. Phys. Lett., **52**, 101 (1988)

9. K. W. Paik, A. L. Ruoff, J. Adhesion Sci. Technol., **2**, 245 (1988)

10. P. Bodo, J.-E. Sundgren, J. Appl. Phys., **60**, 1161 (1986)

11. T. Venkatesan, L. Calcagno, B. S. Elman, G. Foti, ref. 2, p. 301

12. T. Venkatesan, W. L. Brown, C. A. Murray, K. J. Marcantonio, B. J. Wilkens, Polym. Eng. Sci., **23**, 931 (1983)

13. L. Calcagno, G. Foti, Appl. Phys. Lett., **47**, 363 (1985)

14. I. Adesida, C. Anderson, E. D. Wolf, J. Vac. Sci. Technol., **B1**, 1182 (1983)

15. T. H. P. Chang, J. Vac. Sci. Technol., **12**, 1271 (1975)

16. T. M. Hall, A. Wagner, L. F. Thompson, J. Appl. Phys., **53**, 3997 (1982)

17. T. Yamazaki, Y. Suzuki, H. Nakata, J. Vac. Sci. Technol., **17**, 1348 (1980)

18. M. L. Kaplan, S. R. Forrest, P. H. Schmidt, T. Venkatesan, J. Appl. Phys., **55**, 732 (1984)

19. T. Venkatesan, S. R. Forrest, M. L. Kaplan, C. A. Murray, P. H. Schmidt, B. J. Wilkens, J. Appl. Phys., **54**, 3150 (1983)

20. J. Bartko, B. O. Hall, K. F. Schoch, Jr., J. Appl. Phys., **59**, 1111 (1986)

21. S. A. Jenekhe, S. J. Tibbets, J. Polym. Sci. Phys. Ed., **26**, 201 (1988)

22. T. Venkatesan, S. R. Forrest, M. L. Kaplan, P. H. Schmidt, C. A. Murray , W. L. Brown, B. J. Wilkens, R. F. Roberts, Jr., H. Shonhorn, J. Appl. Phys., **56**, 2778 (1984)

23. J. R. Kulish, H. Franke, A. Singh, R. A. Lessard, E. J. Knystautas, J. Appl. Phys., **63**, 2517 (1988)

24. J. R. Stevenson, H. Solnick-Legg, K. O. Legg, in "Biomedical Materials and Devices", J. S. Hanker and B. L. Giammara Eds., Mater. Res. Soc. Symp. Proc., **110**, Pittsburgh, 1989

25. Y. Suzuki, M. Kusakabe, M. Iwaki, M. Suzuki, "Interfaces between Polymers, Metals and Ceramics", Mater. Res. Soc.

Symp. Proceed., to be published

26. A. Ozello, D. S. Grummon, "Interfaces between Polymers,
    Metals and Ceramics", Mater. Res. Soc. Symp. Proceed.,
    to be published

UV LASERS AND POLYMERS

E. Occhiello, F. Garbassi

Istituto Guido Donegani S.p.A.
Via G. Fauser 4
28100 Novara ITALY

Abstract

An overview of how UV lasers have been used to treat polymers is presented. Two main approaches have been followed. By far the most popular has been ablation. Other problems which have been addressed are related to polymerization due to the formation of free radicals by irradiation and to the chemical modification of polymers in selected areas.

Introduction

UV lasers became popular in the early eighties, they include nitrogen lasers, frequency doubled argon ion lasers, frequency quadrupled YAG, etc.. Most frequent nowadays is the use of excimer lasers, based on excited dimers made by a halogen (F or Cl) and an inert gas (Ar, Kr or Xe). Particularly desirable is their capability of providing intense pulses at a very high power. A shortcoming of early generations has been the limited durability of the gaseous mixture which had to be changed fairly often, more recent lasers are much better in this respect.

Since the early eighties a wide body of literature has appeared, favored mainly by the drive to larger scale integration in chips. The use of excimer lasers has been rather extensive in ablation of polymeric, inorganic and metallic materials, the drive being the possibility of a direct-write lithography. Other typical uses for inorganic materials have been for annealing ion-implanted semiconductors or anyway favoring mixing or diffusion. General reviews of laser interactions with materials can be found for instance in (1-3).

In the case of polymers, excimer lasers have been and are considered attractive short-wavelength sources for lithography, as described in other papers in this book. The application which is most frequently met in the literature is by far ablation. Other interesting applications are synthetic. Promoting the formation of free radicals by laser irradiation has been used for polymerization. The directionality of lasers has been used for selective chemical modification of polymers to alter conduction and optical properties of polymers.

115

F. Garbassi and E. Occhiello (eds.), High Energy Density Technologies in Materials Science, 115–122.

Ablative applications

The interest in ablating polymers using excimer lasers arose in the early eighties (4). The main feature, as compared to conventional techniques for polymer ablation, including the use of visible and infrared lasers, was the possibility of drilling clean and well defined holes without charring and burning. A review of most studies in polymer ablation prior to 1986 can be found in (5).

The reason for obtaining clean and precise holes after laser irradiation of polymers is that at low fluences (defined as energy per unit area) the so-called Ablative PhotoDecomposition (APD) occurs (6). As shown in Fig. 1-a, the energy is provided photochemically to the macromolecules, bonds are broken and the emission of ablation products occurs cleanly due to the increase in molar volume. No significant transformation into thermal energy occurs. If, on the other hand, photochemical energy is transformed into heat (vibrational energy), again bond breaking occurs, but ablation is more similar to an evaporation, the kinetic energy is also transmitted to adiacent molecules, inducing the formation of an ill-defined pit (Fig. 1-b).

Most literature evidence agrees on the existence of a threshold of excimer laser fluence (dependent on the material and on the laser wavelength) below which no significant etching occurs. Above the threshold until recently the common belief was that each pulse ablates the same amount of material according to the following empirical relation (5, 7-9):

$$d = m \ln (\Phi / \Phi_{tr})$$

where d is the amount of material etched by each pulse, m is the slope of the etch depth vs. ln (Φ) plot, Φ is the fluence and Φ_{tr} is the threshold fluence. This equation has been also derived theoretically on the basis of polymer degradation kinetics (10),

More recently the importance of thermal contributions has been restated, showing that it can be rather significant, inducing non-linearities in etch depth vs. ln (Φ) plots (11) or even suggesting that d is actually proportional to Φ rather than to ln (Φ) (12).

A general agreement is that the conversion of photochemical energy into thermal energy has the following efficiency order: 350 nm > 248 nm > 193 nm. Davis et al. (13) etched a polymeric resist varying the excimer laser wavelength and remaining close to the threshold fluence. They found a purely photochemical etching only when operating at 193 nm, while they suggested the onset of increasing thermal effects at wavelengths higher than 222 nm.

A large number of studies have been devoted to the analysis of laser ablation products which have been found wavelength and fluence dependent. An example is PMMA (polymethylmethacrylate), the study of etch product distributions showed that while at 193 nm 18% is methylmethacrylate (MMA), at 248 nm only 1% MMA is left, the majority of etch products being larger polymer fragments, due to the lower efficiency of the 248 nm radiation in

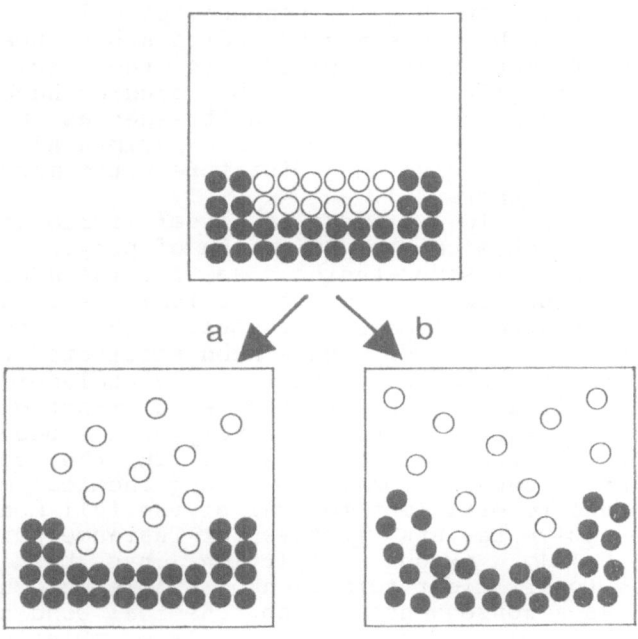

Fig. 1. Ablation mechanisms: a) photochemical, b) thermal.

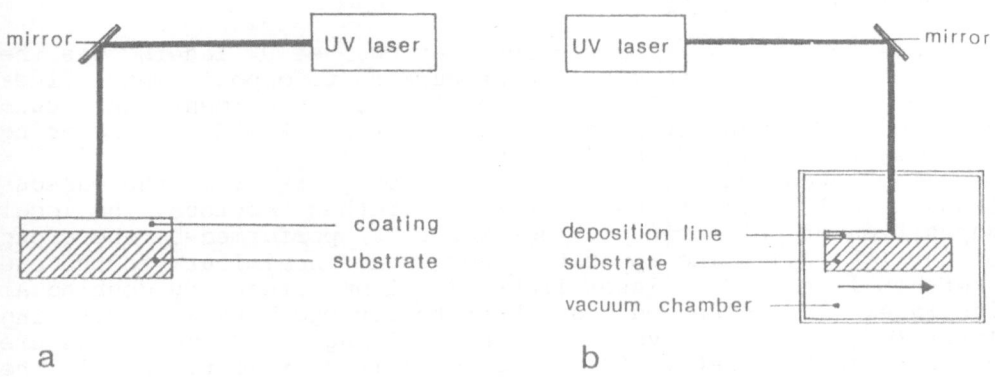

Fig. 2. Schematic apparatus for laser-induced (a) adhesive
         curing and (b) local polymer deposition

bond cleavage (5). At high fluences optical emission from fragments such as carbon ions and CH radicals has been observed, suggesting the formation of a plasma in the vicinity of the irradiated surface (14). A number of studies have also been devoted to the generation of heat and stresses as a consequence of excimer laser ablation. A number of experimental setups have been suggested, including miniature thermocouples and piezoelectric transducers (5).

An interesting side effect of laser irradiation is the morphological and chemical modification of polymer surfaces in the irradiated area. Some studies, see for instance (15), have been devoted to the occurrence of conical structure in many excimer laser ablated polymers, resembling those found in ion bombarded metal surfaces. These have been attributed to shielding effects by microparticulates present in the polymer, protecting the underlying polymer from ablation by a diffraction effect.

The chemical modification of the polymer has been found very much dependent on the polymer itself. In the case of good positive resists, such as PMMA, no strong chemical modification was observed, due to efficient removal of the full monomeric unit (16). In insensitive resists, such as for instance polyimides and polyesters (16-17), oxygen depletion has been frequently observed, due to efficient formation of stable molecules such as carbon oxides as a consequence of photochemical bond cleavage.

As to the application of excimer laser ablation, this is foreseen mainly in microelectronic processing, even if the speed of excimer laser ablation, as compared to conventional resist patterning techniques, has not allowed so far extensive applications (5). Further interest seems involved in excimer laser-enhanced plasma etching, which might become an interesting combination of different high energy density technologies (18). Finally, applications are envisaged in tissue ablation for surgical purposes (5-6).

Synthetic applications

One of the very early applications of UV lasers was the decomposition of organometallic precursors to deposit metal lines on semiconductors (3, 19-20). After initial experiments in vacuum the deposition was obtained also in solution (21-22) and reducing metal salts complexed in a polymer host (23).

The extension to deposition of polymers from the gaseous phase to all sorts of substrates was rather obvious. The local deposition of polymeric coatings was first performed by the group of Ehrlich (24). The apparatus they used is similar to the one sketched in Fig. 2-b, laser light (257.2 nm, frequency doubled Ar ion laser) was shined onto a substrate covered with a sensitizing layer $(Cd(CH_3)_2)$ in a vacuum chamber holding a 1-30 Torr pressure of methylmethacrylate. PMMA was deposited selectively in the areas exposed to the laser beam.

Another example of polymerization in gaseous phase is the application of excimer lasers to the high pressure polymerization of ethylene (25). High conversions have been obtained and the quantum efficiency was shown to be rather high, approaching 9,000 ethylene molecules polymerized by one absorbed KrF (248 nm) laser photon.

The obvious alternative the curing of polymers in liquid, solid or gel conditions. Curing using UV irradiation is well known to induce polymerization of a number of monomers, actually it is widely used in curing inks, coatings and adhesive layers (26-27). It is therefore immediate to conceive utilizing UV lasers, which offer advantages such as high intensity and spectral coherence and the advantage/disadvantage of directionality. The apparatus is similar to that in Fig. 2-a, with the laser light being shined onto the polymerizing medium on a substrate.

Many examples exist both in papers and in patents, using cw UV and excimer lasers. Due to the susceptibility to polymerization by UV irradiation, acrylates and acrylamides have been by far most popular (28-29). Other vinyl-group containing polymers have been polymerized (30), including fumaronitrile - vinylnaphtalene systems (31).

In an interesting study about polymerization of epoxy-acrylate systems (32) the application of pulsed nitrogen lasers was shown to be particularly efficient in initiating the polymerization of multiacrylate resins. Exposure for a few tens of nanoseconds were sufficient to effect complete insolubilization of epoxy -acrylate coatings.

Comparative studies of the relative efficiency of ultraviolet/visible and UV laser irradiation in the photon-initiated polymerization of cyclohexeneoxide/maleic anhydride systems (33) showed that the laser-initiating procedure is 100 or more times more efficient.

Other chemical spatially-selective modifications of polymers have been performed using UV lasers. Among them it is worthwhile remembering the formation of conductive paths in polyimide coatings by graphitization (34). The same experiment, e.g. direct writing of opaque and conductive paths, was also performed on plasma C-H polymers (35-36).

Another very interesting application of UV and visible lasers is in photochemical hole burning for optical storage (37-41). The principle is to irradiate a solid transparent polymer (usually PMMA) containing photosensitive molecules, e.g. tetrazines (38), polyacenes (40) or porphirins (41). These molecules upon irradiation undergo chemical changes which induce the formation of small areas with different optical properties.

Conclusion

The use of UV lasers in treating polymers has aroused quite a lot of interest. Ablation by excimer lasers and/or their use for lithography is ready for technological application, even if an increase in processing speed is necessary. Other high-tech applications such as hole-burning for optical storage information are also very interesting. In mass production applications, some promise is shown by applications to curing of coatings and adhesives.

References

1. T. J. Chuang, J. Vac. Sci. Technol., $\underline{21}$, 798 (1982)

2. D. J. Ehrlich, J. Y. Tsao, J. Vac. Sci. Technol., $\underline{B1}$, 969 (1983)

3. B. R. Appleton, B. Sartwell, P. S. Peercy, R. Schaefer, R. Osgood, Mater. Sci. Engineer., $\underline{70}$, 23 (1985)

4. R. Srinivasan, W. J. Leigh, J. Am. Chem. Soc., $\underline{104}$, 6784 (1982)

5. J. T. C. Yeh, J. Vac. Sci. Technol., $\underline{A4}$, 653 (1986)

6. B. J. Garrison, R. Srinivasan, J. Appl. Phys., $\underline{57}$, 2909 (1985)

7. J. H. Brannon, J. R. Lankard, A. I. Baise, F. Burns, J. Kaufman, U.S. Patent No. 4,508,749 (1985)

8. P. E. Dyer, J. Sidhu, J. Appl. Phys., $\underline{57}$, 1420 (1985)

9. T. F. Deutsch, M. W. Geis, J. Appl. Phys., $\underline{54}$, 7201 (1983)

10. H. H. G. Jellinek, R. Srinivasan, J. Phys. Chem., $\underline{88}$, 3048 (1984)

11. V. Srinivasan, M. A. Smrtic, S. V. Babu, J. Appl. Phys., $\underline{59}$, 3861 (1986)

12. G. D. Mahan, H. S. Cole, Y. S. Liu, H. R. Philipp, Appl. Phys. Lett., $\underline{53}$, 2377 (1988)

13. G. H. Davis, M. C. Gower, Appl. Phys. Lett. $\underline{50}$, 1286 (1987)

14. G. Koren, J. T. C. Yeh, J. Appl. Phys., $\underline{56}$, 2120 (1984)

15. P. E. Dyer, S. D. Jenkins, J. Sidhu, Appl. Phys. Lett., $\underline{49}$, 453 (1986)

16. R. Srinivasan, S. Lazare, Polymer, $\underline{26}$, 1297 (1985)

17. R. Srinivasan, S. Lazare, J. Phys. Chem, $\underline{90}$, 2124 (1986)

18. W. Holber, D. Gaines, C. F. Yu, R. M. Osgood, Jr., Appl. Phys. Lett., $\underline{52}$, 1204 (1988)

19. D. J. Ehrlich, R. M. Osgood, Jr., T. F. Deutsch, J. Vac. Sci. Technol., 21, 23 (1982)

20. G. S. Higashi, L. J. Rothberg, J. Vac. Sci. Technol., B3, 1460 (1985)

21. R. J. von Gutfeld, M. H. Gelchinski, L. T. Romankiw, J. Electrochem. Soc., 130, 1840 (1983)

22. R. K. Montgomery, T. D. Mantei, Appl. Phys. Lett., 48, 493 (1986)

23. A. Auerbach, J. Electrochem. Soc., 132, 1437 (1985)

24. J. Y. Tsao, D. J. Ehrlich, Appl. Phys. Lett., 42, 997 (1983)

25. M. Buback, H.-P. Voegele, Makromol. Chem., Rapid Commun., 6, 481 (1885)

26. P. Pappas, "UV Curing Science and Technology", Technology Marketing Corporation, Stamford, CT, 1978

27. G. E. Green, B. P. Stark, S. A. Zahir, J. Macromol. Sci. Rev. Macromol. Chem., C21, 187 (1982)

28. L. P. Parts, W. R. Feairheller, Jr., U.S. Patent No. 3,477,932 (1969)

29. C. Decker, Polym. Photochemistry, 3, 131 (1983)

30. R. K. Sadhir, J. D. B. Smith, P. M. Castle, J. Polym. Sci. Chem. Ed., 21, 1315 (1983)

31. M. A. Williamson, J. D. B. Smith, P. M. Castle, R. N. Kaufmann, J. Polym. Sci. Phys. Ed., 20, 1875 (1982)

32. C. Decker, J. Polym. Sci. Chem. Ed., 21, 2451 (1983)

33. R. K. Sadhir, J. D. B. Smith, P. M. Castle, J. Polym. Sci. Chem. Ed., 23, 411 (1985)

34. J. I. Raffel, J. F. Freidin, G. H. Chapman, Appl. Phys. Lett., 42, 705 (1983)

35. S. Prawer, R. Kalish, M. Adel, Appl. Phys. Lett., 48, 1525 (1986)

36. S. Prawer, R. Kalish, M. Adel, Appl. Phys. Lett., 49, 1157 (1986)

37. G. Castro, D. Haarer, R. M. Macfarlane, H. P. Trommsdorff, U.S. Patent No. 4,101,976 (1978)

38. E. Cuellar, G. Castro, Chem. Phys., 54, 217 (1981)

39. J. Friedrich, D. Haarer, Angew. Chem. Int. Ed. Eng.,
    23, 113 (1984)

40. M. Iannone, G. W. Scott, D. Brinza, D. R. Coulter,
    J. Chem. Phys., 85, 4863 (1986)

41. W. E. Moerner, T. P. Carter, C. Brauchle, Appl. Phys. Lett.,
    50, 430 (1987)

Part III

CONTRIBUTED PAPERS

KINETIC ASPECT OF PACVD MODELING

Maurizio Masi[1], Massimo Morbidelli[2] and Sergio Carrà [1]

(1)   Dipartimento di Chimica Fisica Applicata,
       Politecnico di Milano,
       Piazza Leonardo da Vinci, 32, 30133 Milano, Italy

(2)   Dipartimento di Ingegneria Chimica e Materiali,
       Università di Cagliari.
       Piazza d'Armi, 09123 Cagliari, Italy

Abstract

A complete model for PACVD reactors involving detailed
calculations of the reaction rate constants,   and the diffusion
coefficients of the neutral and free radical species, based on
the ambipolar approximation for glow discharges,   is applied to
the deposition of silicon nitride from silane and ammonia. the
model results are compared with experimental data   obtained by an
Advanced Semiconductor Materials reactor named Plasma Enhanced
CVD III.

Introduction

Plasma Assisted Chemical Vapor Deposition (PACVD) reactors
are used in the microelectronic industry for the   deposition and
the etching of thin solid films (1.2). These reactors are
constituted by a discharge in a weakly ionized gas,   which is
sustained by an external   electric field powered by a continuous
(DC) or a radio-frequency (RF) generator.

One of the main characteristic of such a discharge is the
high difference between the two species (3). As a consequence the
discharge is not at local equilibrium,   so that the highly
energetic electrons ate able to   ionizing and dissociating the
neutral species at high   rates even though   the bulk gas
temperature is quite low.

The modelling of these reactors implies a   detailed analysis
of the following chemico-physical aspects:

a -   evaluation of the rate constants of the   involved elementary
       gas phase excitation, ionization and dissociation reactions;

b -   description of the diffusional processes of the significant
       radical species from the bulk fluid phase to the deposition
       surface;

c -   kinetics of the deposition reactions.

125

F. Garbassi and E. Occhiello (eds.), High Energy Density Technologies in Materials Science, 125–131.
© 1990 *Kluwer Academic Publishers.*

In the following the above mentioned aspects are analysed for the deposition of silicon nitride from silane and ammonia. In particular, the role of some operative parameters, such as pressure and applied power, on the deposition rate is investigated.

Plasma modelling

After the calculation of Graves and Jensen (4) the ambipolar approximation for describing the discharges is adopted (3,5), so the following equations in terms of electron concentration and electronic temperature:

$$\frac{\partial n}{\partial t} - \nabla \cdot (D_a \nabla n) = n \sum_{j=1}^{NS} k_{ij} n_j \tag{1}$$

$$\frac{\partial T_e}{\partial t} = -\chi_{ea} \nu_{ea} (T_e - T_a) + \frac{2}{3} \cdot \frac{e^2 E^2}{m \nu_{ea}} \tag{2}$$

with n=0 as the boundary condition for eq. (1), at both the electrode surfaces.

Eq. (2) is obtained by assuming that all the energy given to the plasma by the applied electric field, is dissipated by the elastic and anelastic collisions of the electrons with the other species which are present in the plasma.

For a parallel reactor, i.e. the reactor mentioned above, the electric field in the plasma can be expressed by the following equation (6):

$$E = \frac{J}{\epsilon_0 \omega} \left[\frac{1 + \alpha^2}{(r-1)^2 + \alpha^2} \right]^{0.5} \tag{3}$$

Where is the frequency of the oscillating electric field, ϵ_0 is the permittivity of vacuum, α and r are dimensionless parameters defined in the notation. The effect of the oscillating voltage applied to the system can be accounted for by introducing an effectiveness electric field as follow (3):

$$E^2 = \frac{E_0^2}{2} \cdot \frac{\nu_{ea}^2}{\nu_{ea}^2 + \omega^2} \tag{4}$$

thus, the system can be considered in pseudo steady-state conditions.

The above equations can be solved by numerical techniques, such as the Galerkin Finite Elements Methods for two- or tri-dimensional reactors, and Orthogonal Collocation Method for monodimensional ones, such as the case considered here. The evaluation of the ambipolar and electron diffusion coefficients is performed through relationship reported in (3), according to the ambipolar theory.

Kinetic Aspects

One of the main characteristics of PACDV reactors is the high number of species and relative elementary reactions involved. The detailed modeling of such a large spectrum of reactions is not convenient in practice, and a simplification of the network of reactions is necessary. In particular, in the case under consideration, the most important reacting species with respect the deposition process are the free radical species. The ionic species can be neglected in the scheme of deposition reactions due to their very low concentration with respect to that of the free radicals (glow discharges exhibit a degree of ionization of about 10^{-7}). Thus, according to the analysis reported by Claassen (7) the following simplified kinetic scheme has been considered:

ionization
$$SiH_4 + e^- \xrightarrow{k_{i1}} > SiH_4^+ + 2\,e^-$$

$$NH_3 + e^- \xrightarrow{k_{i2}} > NH_3^+ + 2\,e^-$$

dissociation
$$SiH_4 + e^- \xrightarrow{k_{d1}} > SiH_2 + H_2 + e^-$$

$$NH_3 + e^- \xrightarrow{k_{d2}} > NH + H_2 + e^-$$

gas phase reaction
$$SiH_4 + SiH_2 \xrightarrow{k_3} > Si_2H_6$$

deposition
$$SiH_2 + x\,NH \xrightarrow{R} > SiN_xH_y + (2 + x - y)/2\,H_2$$

where the ionization reactions are only responsable, for sustaining the plasma, while the other reactions, involving electrons, neutral and free radical species affect the deposition process. Due to the low temperature of the gas phase and to the high values of the reaction rate constants, which do not depend upon the temperature T_a of the gas but on the much higher electronic temperature T_e of the plasma, the mass transport of SiH_2 from the bulk gas phase to the deposition surface is assumed as the rate determining step of the deposition process. Assuming pseudo-steady-state conditions, we obtain the following expression for the deposition rate (8).

$$R = \frac{ak_c k_{d1} n \cdot n_{SiH_4}}{k_3 n_{SiH_4} + ak_c} \tag{5}$$

Evaluation of the reaction rate constants

All the reactions considered in the kinetic scheme above for the deposition of Si_3N_4 from SiH_4 and NH_3 are bimolecular . Thus, the correspondent rate constants can be evaluated by means of the theory of bimolecular reactions as reported by Eliason and Hirschfelder (9):

$$k = \frac{8\pi}{m^2} \int_0^\infty \epsilon \cdot \sigma(\epsilon) \cdot f(\epsilon) \, d\epsilon \tag{6}$$

Two quantities are important for the reliability of the model prediction: the electron energy distribution function, $f(\epsilon)$ and the reaction cross section, $\sigma(\epsilon)$. In The case of reactions which do not involve electrons, the reacting species are at the same characteristic temperature, i.e. the temperature of the gas T_a (3,4), and the molecular energy distribution differs significantly from the maxwellian law, leading to the Arrhenius form of reaction rate constant. On the contrary, when electrons are involved, the electron energy function can be obtained by solving the Fokker-Planck equation' as shown in (3,10). Generally, at glow discharge conditions, only the isotropic part of the solution is important, which can be approximated by the Druyvesteyn distribution (3,11).

Table 1: Ionization and dissociation potentials		
Reaction	Compound	Potential (eV)
Ionization	SiH_4	12.3
	NH_3	10.2
	H_2	15.4
Dissociation	SiH_4	8.0
	NH_3	5.9

The reaction cross section can be calculated through some semi-empirical equations, such as the ones reported in (3,12). In this work the Drawin formula has been applied (13). The adopted values for the ionization and dissociation potentials are reported in Tab.1.

Mass transfer coefficient estimation

The estimation of mass transfer coefficient can be performed by usual relationships. The fluodinamic regime of the reactor, in the conditions considered in this work, is laminar, characterized by a Reynolds number, Re equal about 300. In the reactor configuration considered, a fully developed flow between two parallel plates is considered, so that the value of the Sherwood number is assumed equal to Sh = 7.6 (14). Thus, the value of the mass transfer coefficient, k_c is given by:

$$k_c = Sh \cdot D/\delta \qquad (7)$$

where δ is the value of the height of the mass transport boundary layer, which is assumed equal to half of the distance between the electrodes.

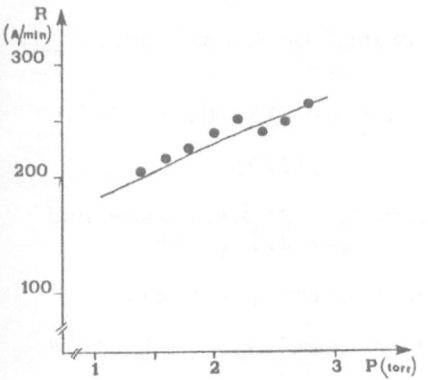

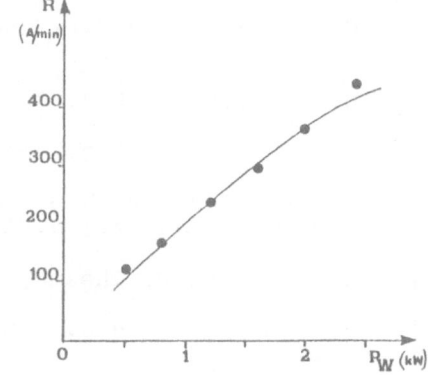

Fig. 1. Variation of
deposition rate
with pressure

Fig. 2. Variation of
deposition rate
with applied power

130

Comparison with the experimental data

In Figs.1 and 2 the comparison between calculated and experimental results (15) is shown for the deposition of silicon nitride from ammonia and silane. The deposition is performed with a reactor which operates at a frequency = 55kHz, whith a residence time, τ = 0.5s and an ammonia silane feed flow rate ratio equal to 8. The obtained agreement is satisfactory particularly when considering that no empirical adjustment of the model parameters has been performed. In particular the effect of pressure and applied power is analized in Fig.1 and Fig.2 respectively.

Conclusions and significance

The use of mathematical models allow the description of the behavior of complex systems such as the one involved in PACVD. In particular, such models allow to explore a wide range of operating conditions for the reactor and then to optimize its performance.

References

1. J.W. Coburn and H.F. Winters, J.Vac.Sci.Technol., 16, 391 (1979)

2. D.L. Flamm and V.M. Donnelly, Plasma Chem and Plasma Proc., 1, 319 (1981)

3. V.E. Golant, A.P. Zilinskij and S.E. Sacharov, Osnovy fiziki plasmy, Mir, Moscow (1983)

4. D.B. Graves and K.F. Jensen, IEEE Trans. on Plasma Sci., 14, 78 (1986).

5. W.P. Allis and D.J. Rose, Physical Review, 93, 84 (1954).

6. A.T. Bell, Ind.Eng.Chem.Fundam., 9, 679 (1970).

7. W.A.P. Claassen, W.G.J.N. Valkemburg, M.F.C. Willemsen and W.M. v.d. Wijgert, J.Eletrochem.Soc.,132, 893 (1985)

8. M. Masi, Ph.D. Thesis, Politecnico di Milano, (1988).

9. M.A. Eliason and J.O. Hirschfelder, J.Chem.Phys., 30, 1426 (1959).

10. H. Margenau, Physical Review, 69, 508, (1946).

11. K. Tachibana, M. Nishida, H. Harima and Y. Urano, J.Phys. D:Appl.Phys., 17,1727 (1984).

12. S. Rhee and J. Szekely, J.Eletrochem.Soc., _133_, 2194 (1986).

13. E. Bauer and C.D. Bartky, J.Chem.Phys., _43_, 2466 (1965).

14. A.H.P. Skelland, _Diffusional Mass Transfer_, Wyley, New York (1974).

15. C. Gatti, Thesis, Politecnico di Milano (1987)

Notation

a	deposition surface per unit volume, cm^{-1}
D	diffusion coefficient, cm^2/s
D_a	ambipolar diffusion coefficient, cm^2/s
e	electric charge of the electron, C
E	intensity of the electric field, V/cm
f	electron energy distribution function
I	ionization potential, eV
J	current density in the discharge, A/cm^2
k_c	mass transfer coefficient of SiH_2, cm/s
k	reaction rate constant
m	mass of the electron, gr
n	electronic concentration, cm^{-3}
n_i	concentration of the i-th component, cm^{-3}
P_w	applied power, W
r	$= ne^2/(m\epsilon_0\omega^2)$
t	time, s
T_a	temperature of the gas, eV
T_e	electronic temperature of the plasma, eV

greek letters

α	$= \nu/\omega$
δ	height of mass transfer boundary layer, cm
ν	frequency of elastic collisions of the electrons, s^{-1}
ν_{ea}	total frequency of collisions of the electrons, s^{-1}
σ	reaction cross section, cm^2
τ	residence time, s
χ_{ea}	energy transfer coefficient
ω	frequency of the oscillating electric field, s^{-1}

subscript

d	dissociation
i	ionization

ADVANCED TECHNOLOGIES FOR ULSI DEVICE PRODUCTION

Giorgio Degiorgis and Francesca Illuzzi

Central R&D, SGS-Thomson Microelectronics
via Olivetti 2, 20041 Agrate Brianza, Italy

Abstract

This paper deals with some oftoday's and tomorrow's high energy density technologies for architecture definition in advanced device production. The present and future status of microelectronics production and of microlithography and plasma etching technologies are briefly outlined.

Introduction

The design rule for device production has shrunk from 10 μm to 0.7 μm in the last decade and the smallest featues are expected to decrease to 0.5 μm in 1990.

As the dimension decreased, device complexity has progressed beyond large scale integration (LSI, 103 - 105 device circuits) through very large scale integration (VLSI, 105 - 106 device circuits) to the ultra large scale integration (USLI, 106 - 107 device circuits). Figure 1 traces the evolution of memory production during the 1980-1985 period. The average feature size decreased from 5.0 μm to 2.5 μm while the minimum teature dimension in leading edge design dropped from 2.0 μm to 0.75 μm.

The complexity and functionality of circuit design has evidently grown with the decreasing of feature size. At the same time alignment requirements among the different layers deceased from 1.0 μm to 0.2 μm.

Increased integration capabilities put a limit problem to reduce cost, so larger wafers were introduced to get more chips per wafer, there is now a trend to 6 in. or even 8 in. diameter wafers (150-200 mm).

The chip size also increased, from 0.5 cm^2 to 'more than 1.5 cm^2, as each generation incorporated smaller components but more functionality. The design rule trends of the microelectronic technologies are plotted in Fig. 2.

Vertical dimensions also decreased. The critical dielectric gate film thickness thinned from 40 nm to 15 nm and the junction depth of the active element structure shrunk from 400 nm to 200 nm or less.

Advanced microlithography

The microlithography process is designed to define a resist pattern allowing to create a structure in the underlying layer by etching or implanting the areas not protected by the photoresist (1-2).

F. Garbassi and E. Occhiello (eds.), High Energy Density Technologies in Materials Science, 133-138.
© 1990 Kluwer Academic Publishers.

134

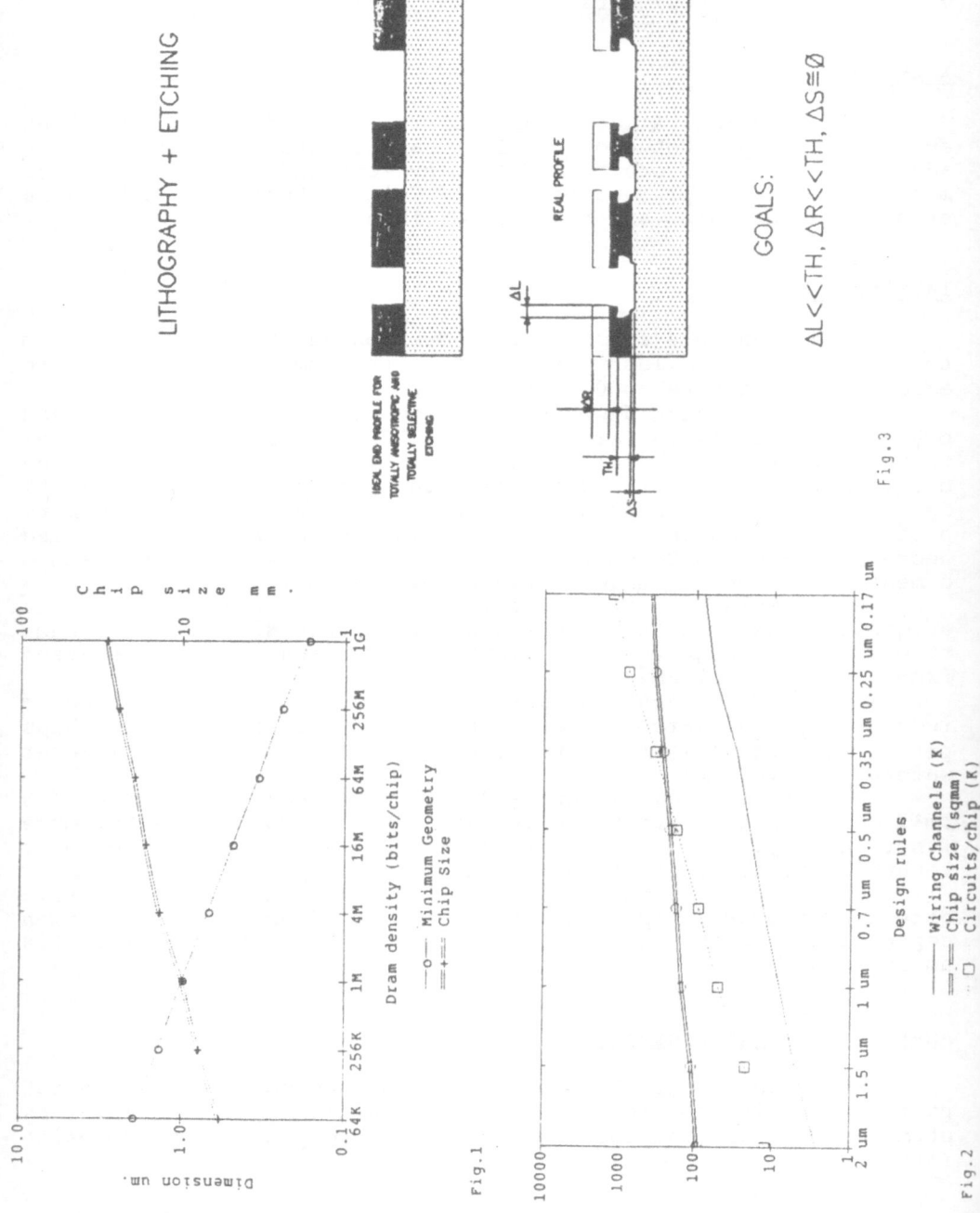

LITHOGRAPHY + ETCHING

GOALS:

$\Delta L \ll TH, \Delta R \ll TH, \Delta S \cong \emptyset$

Fig.3

Fig.1

Fig.2

In projection lithography the image is formed in the resist layer through a UV ligh source, a master mask and a lens system. Presently three wavelengths of the mercury lamp spectrum are normally used for resist exposure: 436 nm (G-line), 405 nm (H-line) and 365 nm (I-line). In good approximation the minimum resolution for projection lithography is given from Raleigh criteria: the resolution is proportional to the wavelength divided by the numerical aperture of the lens. Because of the difficulty to design a lens system with good performance in the MUV and DUV range, it seems easier to make lenses with high numerical aperture, but it becomes difficult to control the focus, since it is equal to wavelength divided by the square of the numerical aperture. Presently production environments use G-line systems with numerical aperture 0.4-0.5. To improve resolution new I-line systems with numerical aperture 0.35-0.45 are already in use in advanced production lines. The use of interferometric laser systems for alignment allows to reach a mask to mask overlay lower than 0.25 µm with very good reproducibility.

The goal of a high resolution patterning could be reached in the future in two different ways: 1) with the standard projection exposure lithography by using DUV light sources, 2) with direct writing using dedicated techniques.

In the first case, the resolution can be improved by using DUV laser sources, providing short wavelengths, high intensity and good spatial coherence. Some advanced labs tested excimer lasers, in particular ArF (193 nm) and KrF (248 nm). The main problem is that it is not possible to use the present resist systems. Good results (0.13 µm) have been reached using very thin "diamon-like" films, the high intensity source photolyzes the surface layer.

Another important source for projection lithography are the X-ray wavelengths, but two important problems have been met in this particular case. The first is the need of a synchrotron to obtain a collimated beam, the second the difficulty to find a good material transparent to X-rays for mask making.

A new interesting technique is the so called "Plasmask" process. Using a thick planarizing layer and exposing only the top of the layer it is possible to reach a high resolution without any defocussing problem. After the exposure step it is then possible to selectively diffuse silicon containing amines into the exposed areas. Thereafter in oxygen RIE (reactive ion etching) or MIE (magnetron ion etching) conditions silicon forms an oxide and protects the exposed regions, while the unexposed ones are etched by the plasma. This process presents a very low defectivity as compared to normal wet development.

The other future approach to high resolution patterning is direct writing. Many different sources may be used for direct writing both in resist layer patterning and in direct etch, film deposition and ion implantation. Right now the main drawback of direct writing technologies is the very low throughput. The advantage is their superior precision and resolution, therefore important improvements are expected in the near future.

Direct writing electron beam (EB) lithography ia already ured since a long time for mask production and its use is expanding into the patterning of special custom designed (ASIC) devices. Similarly ion beam (IB) sources may be used, but the possible damage of underlying layers due to the high mass of ions has yet to be solved.

In future the use of laser sources will probably very important, since they can be used for different processing steps, varying wavelength and power emission.

Plasma etching technology

The only methodology allowing pattern transfer with high precision and fast throughput is based on plasma etching (3-5). In Tab. I the main dry etching processes utilized for 1-Mbit EPROM production are reported.

In order to satisfy to shrunk design rules in Fig. 2, plasma etching must provide good uniformities (uniformity will become more and more important for large diameter wafers), high selectivity (the gate oxide thickness becomes thinner and thinner), high directionality (to preserve line sizes) and good reproducibility. In Fig. 3 the conditions for anisotropic and selective etching are shown.

Unfortunately there is a contrast between uniformity, anisotropy and selectivity for most etching processes. The etch parameters enhancing anisotropy (density of ions) have a negative effect on selectivity (enhanced by high active species density). Furthermore optimizing surface topography requires long over-etch times, which in turn decrease both selectivity and anisotropy.

In order to develop a process satisfying the previous requirements, the best compromise must be found. For instance a multistep process can be developed optimizing anisotropy and selectivity at each step.

An important problem of plasma etching processes is device damage. During a plasma discharge, energetic particles impinge on the wafer surface, these particles may cause alterations in device performace. Damages can arise from:

1) Fluorocarbon ion bombardment producing fluorinated surface layers, changing the layer resistivity.

2) Proton penetration in the substrates, when hydrogen containing gas mixtures are used.

3) Charge accumulation in the gate oxide.

4) Unwanted impurity that contaminate the metal layers.

5) Bombardment with high energy photons.

Future efforts in plasma etching processes will have to be devoted to damage control and selectivity improvement (for thinner substrates and higher etch rate materials). Works are in

TABLE I		
LAYER	*STRUCTURE*	*ETCH REQUIREMENTS*
Si_3N_4	Field oxide	Anisotropy Selectivity to thin oxide
WSi_2/polysilicon	Gate line	Linesize control No residues High selectivity to oxide Undercut control
Oxide	Contact windows	Anisotropy Selectivity to silicon Minimal polymer growth
Aluminum	Interconnects	Anisotropy No corrosion
Photoresist	All patterning and implant mask	No damage

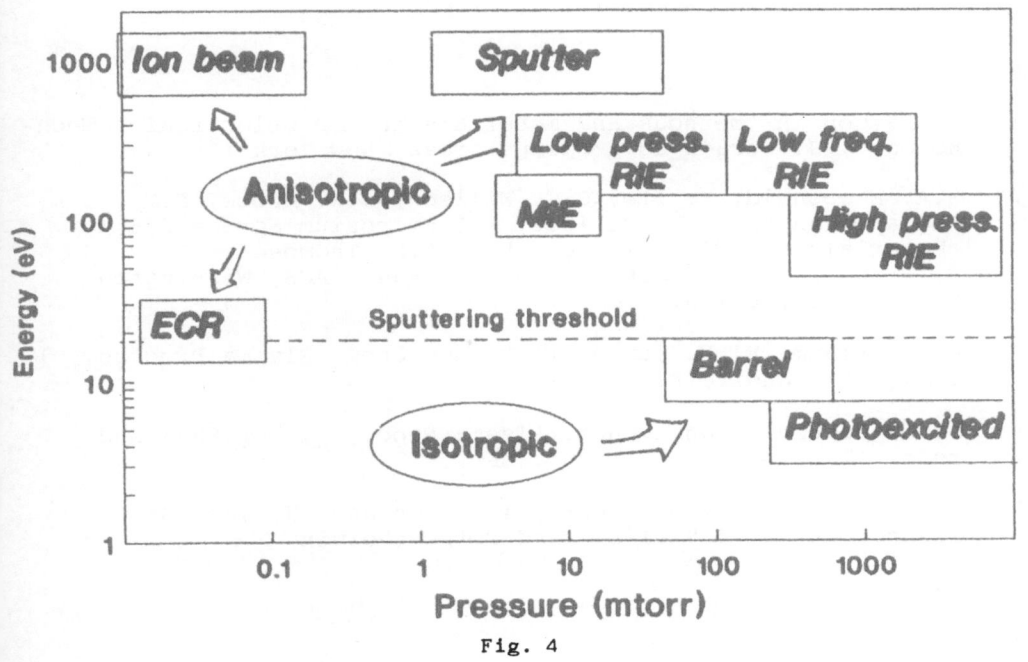

Fig. 4

138

progress all over the world in order to reduce energetic bombardment and to increase at the same time etchant density. In Fig. 4 the relationship between ion energy and etching pressure is reported for different etching systems.

For instance the MIE configuration produces increased etchant density, in fact the electrons are concentrated in proximity of the wafer and create higher etchant density by elastic collisions with gas molecules. At the same time the wafer is submitted to lower bias voltages and therefore to lowre damage. Furthermore in magnetron systems the wafer is surrounded by a "magnetic bottle" insulating it from reactor walls thus minimizing contamination.

Another system which has recently attracted lots of attention is ECR (electron cyclotron resonance. In a ECR configuration microwave energy (2.45 GHx)is coupled into a waveguide and transmitted to the etching chamber, within a magnetron. Fig. 4 shows the low wafer bombardment in ECR systems. Unfortunately etch rates are rather low in ECR systems, therefore the implementation of ECR machines in ULSI production is questionable.

The triode configuration itself can solve some problems. The separation of power supplies, for instance between grid or wall and unloaded electrode may control ion bombardment to the wafer loaded electrode and allow etchant selection at the wafer surface.

A last interesting etching technique is excimer laser assisted etching. In this case the energy normally supplied by ions is introduced by laser photons. The laser beam directed on the wafer surface assists the etching reaction allowing a high degree of anisotropy and low damage.

References

1. J. Bargon in "Methods and Materials in Microelectronic Technology", J. Bargon ed., Plenum Press, New York (1984).

2. S.A. MacDonald, H. Ito, C.G. Willson, J.W. Moore, H.M. Gharapetian and J.E. Guillet, ACS Symposium Series 266, "Materials for Microlithography", L.F. Thompson, C.G. Willson and J.M.J. Frechet, Eds., ACS, Washington, D.C. (1984), p.179.

3. D. L. Flamm, V. M. Donnelly, Plasma Chem. Plasma Process., 1, 317 (1981) and refs. therein

4. J. W. Coburn, Plasma Chem. Plasma Proc., 2, 1 (1982) and refs. therein

5. H. F. Winters, J. W. Coburn, T. J. Chuang, J. Vac. Sci. Technol. B, 1, 469 (1983) and refs. therein

DESIGN/OPERATING PARAMETERS OPTIMIZATION OF AN ELECTRON ACCELERATOR FOR SURFACE TREATMENTS ON MATERIALS THROUGH IRRADIATION

G. Matticari

Proel Tecnologie S.p.a. - Ferranti International Signal
Viale Macchiavelli, 29 - Firenze

Abstract

The aim of the paper is to present an analytical procedure for the evaluation and optimization of the functional parameters of an electron accelerator. This accelerator is conceived for the irradiation of surface coatings laid on substrates of different materials, for obtaining a real time curing (for example polymerization). The goal is to maximize the process efficiency on the basis of the coating physical characteristics (density, thickness, curing dose).

Introduction

The general design procedure of an electron accelerator for surface industrial treatment foresees the utilization of both a theoretical simulation and an experimental verification. The theoretical simulation concerns, on the one hand, the electron trajectories evaluation, with the purpose of verifying if a preselected gun configuration satisfies the requested specifications and, on the other hand, the process simulation.

We shall present in the following, a possible procedure of process simulation for the theoretical determination of the operative parameters (i.e. accelerating voltage, beam current and irradiation time) of an accelerator, with the goal of maximizing the process efficiency, defined as the ratio between the useful energy transferred to the coating and the total energy of the electron beam.

The analytical procedure show the possibility to maximize the above said efficiency through a suitable choice of the ratio between the curing dose (supposed to be known) necessary to obtain the requested chemical-physical modifications in the coating material, and the peak absorbed dose.

The study we present has been developed assuming a parabolic model for the curve which provides the energy absorption versus the electron penetration and limiting the operation range of the accelerator between 100 and 400 KV.

We foresee to verify the results obtained in this theoretical simulation of the process with a suitable experimental activity that will be carried on in the field of electron accelerator for surface treatments.

F. Garbassi and E. Occhiello (eds.), High Energy Density Technologies in Materials Science, 139–145.
© 1990 Kluwer Academic Publishers.

Terminology and Definitions

The absorbed dose by an object exposed to the action of an electron beam is defined as the energy transferred for unit mass of the irradiated material.

The absorbed dose is expressed in rad, 1 rad = 100 erg/g and consequently:

Mrad = 10 J/g

Therefore: absorbed dose D(Mrad) = 100 Energy(J)/Mass(mg)

We define the mass thickness r the quantity:

$r(mg/cm^2) = \varrho(mg/cm^3) \cdot z(cm)$

Where ϱ is the mass density and z the penetration.

The electron range z_o (cm) is the depth of penetration corresponding to the total loss of energy of the impinging electrons and consequently we have:

Maximum mass thickness $r_o(mg/cm^2) = \varrho(mg/cm^3) \cdot z_o(cm)$

Theoretical Models

The present study has been developed assuming a theoretical model both for the electron penetration as function of the accelerating voltage and for the energy absorption as function of mass thickness.

For the maximum mass thickness r_o (mg/cm^2) we assume:

(1) $r_o = 0.011 \; U^{1.58}$

with U(KV) electron accelerating voltage.

This functional relation has been obtained from interpolation of the experimental law in the range 100÷400 KV. For the absorbed dose D we assume:

$$(2) \quad D(r) = D_{max} \left[1 - \frac{9}{4} \left(\frac{r}{r_o} - \frac{1}{3} \right)^2 \right]$$

This parabolic model of the absorbed dose D as function of mass thickness r is a 2nd order approximation of the experimental law in the range 100÷400 KV.

The quantity D_{max} corresponds to the peak absorbed dose and can be easily obtained through the following relation of energetic balance:

$$\int_0^{r_o} D(r)dr = 100 \; Uj\tau$$

where j (mA/cm²) is electron current density, τ (s) is the irradiation time and consequently $Uj\tau$ is the total beam energy per
unit area.

Calculating the integral we obtain:

(3) $\quad D_{max} = 100 \cdot \dfrac{4}{3} \dfrac{Uj\tau}{r_o}$

The accelerating voltage U (KV) and the current density
j(mA/cm²) characterize the electron beam upstream the thin metal
foil that separates the high vacuum zone from the zone, filled
with inert gas, interposed between the same foil and the
treatment plane.

It is convenient, at this point, to show the graphic
representation of the adsorbed dose D(r) as a function of the
mass thickness r. In Figure 1 the shadowed areas represent the
different shares of energy absorbed by the different materials.

In Fig. 1 the following notations are used:

r_w = mass thickness of the metal foil.

r_g = mass thickness of the inert gas between the foil and the
product plane.

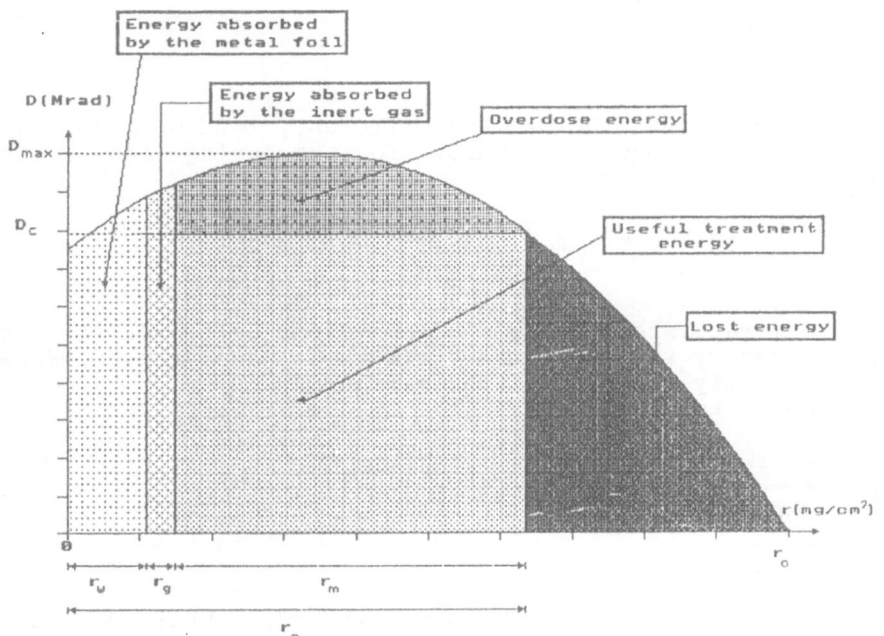

Fig. 1 - Display of dose profile and energetic balance.

r_m = the mass thickness of the coating to be cured through irradiation.

$r_p = r_w + r_g \, r_m$.

D_c (Mrad) = required curing dose.

The condition to be satisfied to obtain the curing of the product is that, within the whole thickness of the coating, the absorbed dose is greater or at least equal to the required curing dose, that is analytically:

(4) $D(r) \geq D_c$ when $r_w + r_g \leq r \leq r_p$

Optimization and final dimensioning of the accelerator

The limit condition we choose for the optimization is:

$D(r_p) = D_c = FR \quad D_{max} \qquad 0 \leq FR \leq 1$

where FR is a certain fraction of the peak dose and is the parameter to be determined.

The above condition brings, as a consequence, to the following relation:

(5) $$\frac{r_p}{r_o} = \frac{1+2\sqrt{(1-FR)}}{3}$$

If we consider the process efficiency η, defined as the ratio between the required energy for the curing and the total energy of the e-beam we have:

$$\eta = \frac{0.01 \, D_c \, r_m}{Uj\tau}$$

The numerator represents the energy per unit area (in J/cm²) That is necessary for the surface treatment and the denominator the same quantity referred to the e-bean.
We may also write:

(6) $$\eta = \frac{D_c \, r_m}{3 \, D_c \, r_o} \Big/ {4 \, FR} = \frac{4}{9} \frac{r_m}{r_p} \, FR \, [1+2\sqrt{(1-FR)}]$$

We can see that the process efficiency, obtained as above, is independent on the curing dose D_c.

The function η (FR) can be easily maximized, obtaining for FR the optimum value of about 0.81.

With the optimum value of FR the relation:

$$(7) \quad \frac{r_w + r_g}{r_w + r_g + r_m} \geq \frac{1 - 2/(1+FR)}{1 + 2/(1+FR)}$$

must be satisfied in order to warrant that the surface dose $D(r_w + r_g)$ is greater or equal to the curing dose Dc. If the optimum value of FR (=0.81) does not satisfy the (7), it is necessary to choose for FR the value

$$FR' = 1 - \frac{1}{4(1 + 2\ r_e/r_m)^2}$$

obtained considering the = in (7) and solving for FR.

With this value of FR all the other parameters of system can be consequently calculated.

At this point we can proceed in the final dimensioning of the e-gun parameters.

The known parameters are: r_w, r_g, r_m (and then r_p).

For FR we take the optimum value $FR_{opt} = 0.81$.

From relation (5) the parameter r_o (maximum mass thickness) can be evaluated. Knowing r_o the necessary accelerating voltage U (KV) can be computed from relation (1).

As the necessary curing dose D_c is supposed known we can obtain the value of the product j (mA/cm²) • τ (s) from the relation (3), therefore:

$$(8) \quad j\tau = \frac{3\ D_c}{400\ FR} \cdot \frac{r_o}{U}$$

In most of the treatment systems τ (irradiation time) is determined by the product feeding speed v (cm/s) and the width of the beam w (cm) along the direction of the product feeding then τ (sec) = w/v. Knowing τ it is possible to calculate the parameter j (mA/cm²) from equation (8).

In case of a computer controlled system U and j can be automatically selected according to the values of parameters r_w, r_g, r_m, D_c, w and v, on the basis of the presented analytical procedure.

Numerical Simulation

Using the analytical procedure shown in the previous paragraph a numerical simulation at the computer has been carried on for some different operation conditions.

The results are summarized in Tab. 1.

EXTRACTION METAL FOIL	Material	Titanium	Aluminium	Titanium	Aluminium	Titanium	Aluminium	Titanium	Aluminium
	Dens.(g/cm3)	4,5	2,7	4,5	2,7	4,5	2,7	4,5	2,7
	Thickness(um)	25	25	25	25	25	25	25	25
INERT GAS IN THE TREATMENT ZONE	Gas	Nitrogenum	Nitrogenum	Nitrogenum	Nitrogenum	Nitrogenum	Nitrogenum	Nitrogenum	Nitrogenum
	Dens.(g/cm3)	0,0013	0,0013	0,0013	0,0013	0,0013	0,0013	0,0013	0,0013
	Thickness(cm)	3	3	3	3	3	3	3	3
SURFACE COATING	Material	Silicone	Silicone	Silicone	Silicone	Epoxy	Epoxy	Epoxy	Epoxy
	Dens.(g/cm3)	1,8	1,8	1,8	1,8	1,5	1,5	1,5	1,5
	Thickness(um)	300	300	300	300	300	300	300	300
CURING DOSE NEEDED FOR THE TREATMENT	Dc (Mrad)	10	10	10	10	5	5	5	5
TOTAL MASS THICKNESS TO PENETRATE	Rp (mg/cm2)	69,15	64,65	69,15	64,65	60,15	55,65	60,15	55,65
RATIO Dc/Dmax	FR	0,8114	0,8114	0,6	0,6	0,8114	0,8114	0,6	0,6
EFFICIENCY OF TREATMENT	$FR(1+2(1-FR)^{0.5})$ $9Rp/4Rm$	52,62%	56,28%	47,17%	50,45%	50,41%	54,49%	45,19%	48,84%
MASS THICNESS CORRESP. TO D(r) = 0	Ro (mg/cm2)	111,02	103,80	91,59	85,63	96,57	89,35	79,67	73,71
ACCELERATING VOLTAGE	U (kV)	342,13	327,86	302,91	290,28	313,23	298,18	277,32	264,00
PRODUCT CURR. DENSITY IRRAD. TIME	j x t (mA.sec/cm2)	3,00E-02	2,93E-02	3,78E-02	3,69E-02	1,42E-02	1,38E-02	1,80E-02	1,75E-02
% OF ABSORBED ENERGY	Metal foil	11,06	6,90	13,61	8,45	12,85	8,08	15,83	9,92
	Inert gas	4,20	4,31	5,22	5,34	4,91	5,09	6,11	6,32

Tab. 1 - Results of the computer simulation.

The obtained results corresponding to the optimum value of FR (=0.81) can be easily compared, for the same operation conditions, with those obtained for a different value of FR (=0.6). The simulation has been performed considering two types of coatings (silicone and epoxy resin), of the same thickness (300 μm).

From the analysis of the data contained in Tab. 1, can be noted that decreasing FR obviously gets worse the efficiency, but decreases the required accelerating voltage for the treatment.

Being the function (FR) rather smooth in the range 0.5÷0.81 could, in same cases, be convenient to choose for FR a value a little lower than the optimum value in order to operate with slightly lower accelerating voltages.

Conclusion

During this brief dissertation we have shown that the working parameters of an electron accelerator for the execution of surface treatment can be optimized with the goal of obtaining the maximum energetic efficiency from the process of coating material irradiation.

We foresee to validate with a suitable experimental activity the results obtained theoretically and to point out, in a second part of paper, the modalities and the results of this experimentation.

References

1. P. Holl - Polymer Physik Gmbh - Beam in on   curing

2. S. Schiller, V. Heising, s. Panzer -   Electron beam technology

3. R.Amboss-Hughes Research Labs, Malibu, California - The design   of large area electron beam guns.

CF$_4$/O$_2$ PLASMA ETCHING OF POLYMERS

E. Occhiello, M. Morra, F. Garbassi

Istituto Guido Donegani S.p.A.
4 via Fauser, 28100 Novara, ITALY

Abstract

CF$_4$/O$_2$ plasmas have been used to etch a variety of polymers (PPE, PS, PMMA, PC). Purely hydrocarbon polymers (PPE and PS) are etched less readily than oxygen containing polymers (PMMA and PC).

The application of bias voltages to polymer samples resulted in an increase of the number and energy of ions impinging on the surface and higher etch rates. A side effect of applying bias voltages was shifting the highest etching efficiency to mixtures richer in CF$_4$ for PS, PC and PMMA, while no change was observed for PPE.

These experimental behaviors are in remarkable agreement with the theoretical explanation of Egitto et al., suggesting the importance of a saturation step in determining the relative etching efficiency of different polymers.

Introduction

In this work we wish to study the effect of positive ion bombardment on etch rates of polymers in CF$_4$/O$_2$ discharges. It is known that ion bombardment induces enhancements in etch rates. Our aim was to find out whether ion bombardment, besides having a physical effect (introducing energy and therefore enhancing etch rates), influences the chemistry of the process as well.

CF$_4$/O$_2$ plasma treatment of polymers has been widely used for dry etching purposes. Its effectiveness is due to the presence of atomic fluorine and oxygen, which react with polymer surfaces to yield volatile carbon oxides and fluorides. Actinometric evidence of oxygen and fluorine in CF$_4$/O$_2$ discharges has been established (5, 6). CF$_4$/O$_2$ discharge etch rate measurements have been performed on PMMA (7-9) and PS (9). Mechanistic studies on CF$_4$/O$_2$ plasma treatment of polymers were carried out by several groups. In particular Egitto et al. tried to relate the etching behavior to the polymer structure (10-11).

We studied the effect of ion bombardment on the CF$_4$/O$_2$ etch rates of some "prototype" polymers. Plasma polyethylene (PPE) was used because its structure is very simple and free of insaturations, even if plasma polymerization induces crosslinking. Polystyrene (PS) and polymethylmethacrylate (PMMA) have been taken as examples of polymers with a hydrocarbon backbone and pendant functional groups, a phenyl ring in PS (vinyl type polymer - negative resist), an ester and a methyl group in PMMA (vinylidene type polymer - positive resist).

147

F. Garbassi and E. Occhiello (eds.), High Energy Density Technologies in Materials Science, 147–153.
© 1990 Kluwer Academic Publishers.

Lastly we dealt with Bisphenol-A polycarbonate (PC) as an example of polymer with aromatic rings and carbonate groups in the backbone.

The samples were mounted on a separately rf-biasable Quartz Crystal Microbalance (QCM), thus the energy of the positive ions bombarding the sample could be precisely controlled independently of the plasma density. Considering that in the apparatus we used the plasma potential is about + 20 V, the maximum ion energy is the sum of the bias voltage plus 20 eV. Most of the ions are expected to have this energy, because the transit time for the ion to traverse the sheath is much longer than the period of the rf (13.56 MHz), so that the ions time average the rf voltages. Furthermore ion-neutral collisions in the sheath are negligible at the operating pressure (10 mTorr) (12-14).

Experimental

The experimental arrangement used in this work has been extensively described elsewhere (12-14). It consists in a parallel plate reactor, the 13.56 MHz radiofrequency is applied through a blocking capacitor to a 15 cm diameter electrode. A silicon dioxide excitation electrode was used throughout the study to prevent plasma polymerization and redeposition of sputtered material on the microbalances.

Etch rates have been measured using two Quartz Crystal Microbalances (QCM) inserted into the glow discharge midway between the electrodes, with the surface located at 10.2 cm from the center of the discharge, the relative orientation is the same but they are 90° apart. The microbalances and holders are from Inficon (Mod. no. IPN 750.040-G1) and water cooled. The bias voltage was applied to one of the QCM holders using a second capacitatively coupled 13.56 MHz rf generator. This bias voltage is induced similarly to the primary electrode (self-bias voltage) and is controlled by the power of the second generator. The second QCM holder was held at ground potential and the etch rates were monitored while the other one was biased to ensure that the bias did not affect the overall glow discharge.

Oxygen and CF_4 (Freon 14) were purchased from Matheson in the form of lecture bottles. In all experiments a 6 cm^3(STP)/min flow of gas was used, the pressure was maintained at 10 mTorr (1.333 Pa) and the discharge power was held at 80 W.

PPE has been deposited by in situ plasma polymerization of ethylene. PS, PMMA and PC were spin cast on the microbalance gold coated quartz crystals from $CHCl_3$ solution and vacuum dryed to remove the solvent. The parent polymers were commercial samples and no attempt was made to extract additives.

Results

The concentration of atomic fluorine in CF_4/O_2 discharges has been shown by actinometry to have a bell-shaped behavior vs. molecular oxygen percentage in the gas mixture, with a maximum at about 30 % oxygen, while the atomic oxygen concentration rises monotonically, in agreement with literature evidence (5-6).

The effect of ion bombardment on etch rates for all polymers was at all gas mixture compositions a nearly-linear increase in etch rate with bias voltage.

The behavior of etch rates as a function of oxygen percentage in the gas feed is shown in Figs. 1-a to 1-d. In the case of PPE an increase in oxygen percentage resulted in enhanced etch rate (Fig. 1-a), at all bias voltages a rather linear increase was observed.

In the case of PS (Fig. 1-b) a steady increase of etch rate with oxygen percentage is observed in the absence of bias voltage. Increasing the latter a bell-shaped form of the curve is developed. At -100 V bias voltage a maximum efficiency of the mixture with 60 % oxygen is apparent. Interestingly the steep increase in etch rate starts at oxygen percentages higher than 20 % at all bias voltages, i.e. close to the maximum fluorine atom concentration in the discharge.

In PMMA (Fig. 1-c) in the absence of bias voltage a sigmoidal curve is observed, with a high etch rate at 20 % oxygen, contrarily to what happened in PS. With increasing bias voltage the efficiency of mixtures with 40 to 60 % oxygen increases and at bias voltages higher than -80 V the 60% oxygen mixture displays maximum efficiency in promoting etching.

PC shows a different behavior (Fig. 1-d). The bell-shaped trend is already apparent in the absence of bias voltage. By increasing the energy of positive ions reaching the surface the etch rate of mixtures with low oxygen percentage are enhanced more effectively than those on the high oxygen percentage side.

Comparing etch rates on different polymers, it is obvious that etching is most efficient on PMMA and PC, while PS and particularly PPE are more resistant. The introduction of oxygen containing groups probably makes available active sites for reactions leading to etching. Furthermore PMMA is a vinylidene type polymer, known to be susceptible to degradation by various physical agents (UV photons, electrons, etc.) (7).

The dependence of etch rates on the gas mixture composition displayed in Figs. 1-a to 1-d suggests that ion bombardment alters the relative rates of reactions leading to etching of PS, PMMA and PC. To gather more information we ratioed the etch rates obtained using different gas mixtures and we obtained constant numbers, independent of bias voltage. This means that even if different reactions are enhanced differently by ion bombardment, the kinetic dependence of etch rates on bias voltage is not changed.

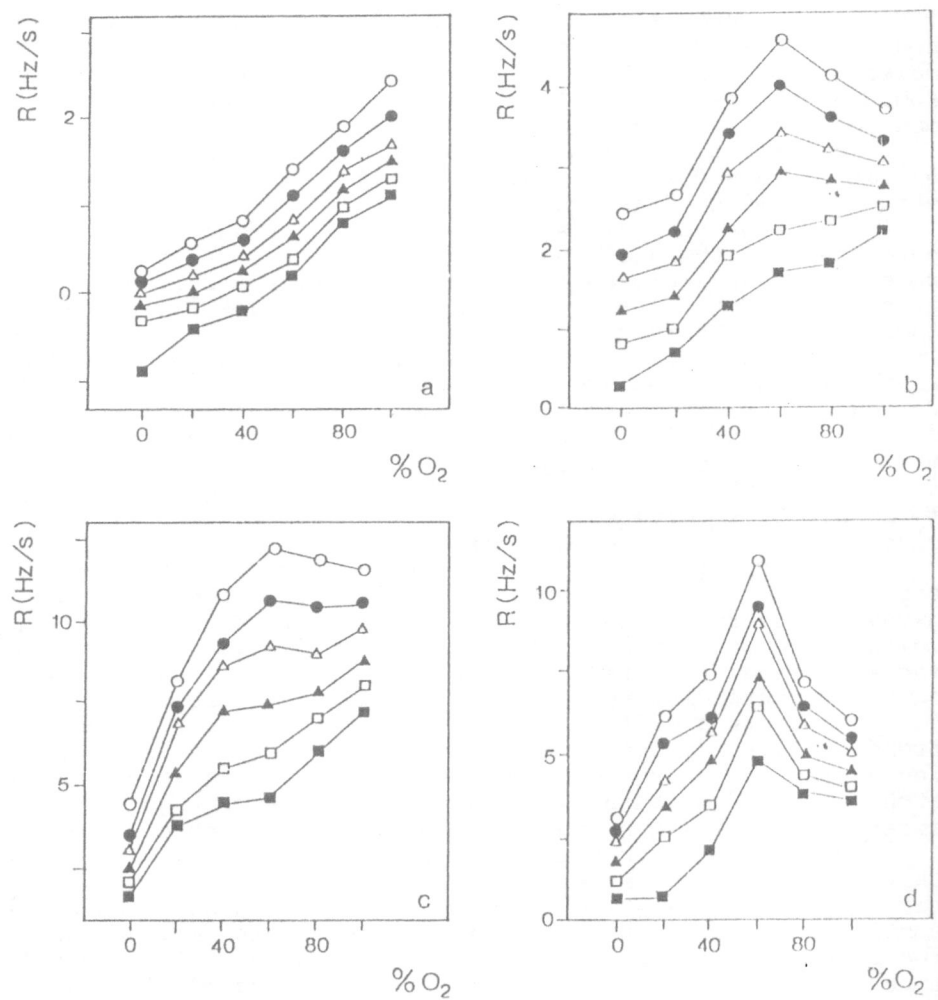

Fig. 1. QCM etch rates by CF_4/O_2 mixtures vs. O_2 percentage varying the bias voltage: 0 V (■), -20 V (□), -40 V (▲), -60 V (△), -80 V (●), -100 V (○). a) PPE, b) PS, c) PMMA, d) PC

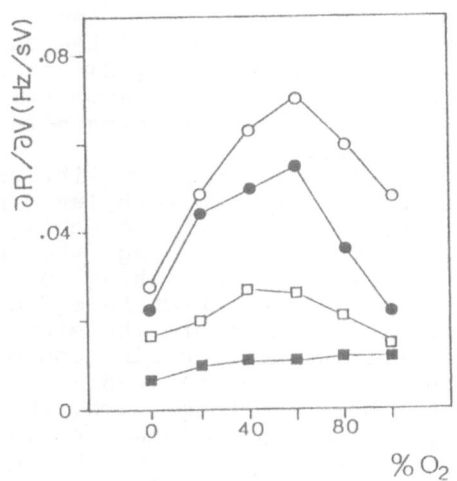

Fig. 2. Relative etching efficiencies as a function of O_2 percentage: PPE (■), PS (□), PC (●), PMMA (o).

Discussion

The efficiency of ion bombardment in enhancing etch rates has been shown to be a function of both the gas mixture composition and the substrate material. But the relative efficiency of different mixtures in promoting etching is altered by ion bombardment. Even if for all mixtures there is a fairly linear increase of etch rate with bias voltage, the enhancement due to ion bombardment is different for each reaction leading to etching, as shown by the fact that the maximum efficiency is shifted to mixtures richer in atomic fluorine for all polymers but PPE. At the same time the kinetic dependence of etch rates on ion bombardment energy is similar for all the mixtures, as shown by the fact that etch rate ratios are independent of bias voltage.

To better exemplify the selectivity of ion bombardment in enhancing etch rates, thanks to the fact that plotting etch rates vs. the bias voltage reasonably linear plots have been found, we used conventional line fitting methods to obtain the corresponding slopes. These slopes can be considered partial derivatives of the etch rate vs. bias voltage. As such they should account for the effect of an increase in number and energy of positive ions impinging on the surface on the etch rate.

In Fig. 2 these slopes, which we named relative etching efficiencies, are plotted vs. the oxygen percentage in CF_4/O_2 mixtures. In PPE there are only very limited changes in relative etching efficiency altering the gas mixture composition, suggesting ion bombardment does not increase selectively the

rate of reactions at the PPE surface. PS and PC have a similar
bell-shaped trend, even if the maximum of the curve rests in one
case (PS) at 40 % O2, in the other (PC) at 60 % O2. In PMMA
again a bell-shaped curve is found, but asymmetrical, since the
relative etching efficiency of pure oxygen is similar to that
of the 20 % O2 CF4/O2 mixture.

Egitto et al. (11-12) performed equilibrium studies of the
etching of polymers (polyimide, polyethylene, polyisoprene) by
CF_4/O_2 discharges. They discussed the molecular orbitals of the
parent and reacted polymers, reaching the conclusion that
fluorine is more effective in attacking unsaturated sites,
leading to saturated intermediates which have to undergo further
reaction to obtain etching. Therefore the etch rate maxima for
saturated polymers occur at higher fluorine concentrations in the
plasma than for unsaturated (aromatic) ones.

By increasing the bias voltage, i.e. the ion energy, the
amount of energy input into the polymer is increased, thereby
moving away the system from an "equilibrium" situation and
therefore reducing the chemical selectivity of the gas mixture.
We observed, in PS, PC and PMMA, a highest relative etching
efficiency for discharges with higher amounts of atomic fluorine
(Fig. 2). That might be, at least for PS and PC, in agreement
with Egitto's results, in the sense that the energy input into
the polymer helps surpassing the suggested saturation step. The
relative indifference of PPE to ion bombardment induces shifts in
relative etching efficiency might also agree with Egitto's
interpretation, since no saturation step is involved (Fig. 1-a).

Acknowledgements

Part of this work was performed at the IBM Almaden Research
Center during a joint research study. The authors wish to express
their gratitude to IBM Corp., to Dr. J. W. Coburn for access to
plasma equipment and helpful suggestions and discussions, to Dr.
E. Gattiglia, for helping in the sample preparation, to Mr. D.
Pearson and Mr. E. Isaacson for technical assistance.

References

1. H. Winters, R. P. H. Chang, C. J. Mogab, J. Evans, J. A.
    Thornton, H. Yasuda, Mater. Sci. Eng., 70, 53 (1985)

2. J. W. Coburn, Plasma Chem. Plasma Process., 2, 1 (1983)

3. H. F. Winters, J. W. Coburn, T. J. Chuang, J. Vac. Sci.
    Techn. B, 1, 469 (1983)

4. D. L. Flamm, V. M. Donnelly, Plasma Chem. Plasma Process., 1,
    317 (1981)

5. J. W. Coburn and M. Chen, J. Appl. Phys., 51, 3134 (1980)

6. R. d'Agostino, F. Cramarossa, S. De Benedictis, G. Ferraro,

J. Appl. Phys., <u>52</u>, 1259 (1981)

7. K. Harada, J. Appl. Polym. Sci., <u>26</u>, 1961 (1981)

8. B. J. Wu, D. W. Hess, D. S. Soong, A. T. Bell, J. Appl.
    Phys., <u>54</u>, 1725 (1983)

9. G. N. Taylor, T. M. Wolf, L. E. Stillwagon, Solid State
    Electron., <u>27</u> (2), 145 (1984)

10. F. D. Egitto, V. Vukanovic, F. Emmi, R. S. Horwath,
    J. Vac. Sci. Technol., <u>B3</u>, 893 (1985)

11. S. R. Cain, F. D. Egitto, F. Emmi, J. Vac. Sci. Technol.,
    <u>A5</u>, 1578 (1987)

12. K. Koehler, J. W. Coburn, D. E. Horne, E. Kay, J. H. Keller,
    J. Appl. Phys., <u>57</u>, 59 (1985)

13. N. C. Us, R. W. Sadowski, J. W. Coburn, Plasma Chem. Plasma
    Process., <u>6</u>, 1 (1986)

14. F. Fracassi, E. Occhiello, J. W. Coburn, J. Appl. Phys., <u>62</u>,
    3980 (1987)

LASER INDUCED SURFACE MODIFICATION OF POLYMERS

V. Malatesta

Istituto Guido Donegani
Via G. Fauser 4
28100 Novara ITALY

Introduction

It has been shown that irradiation with ultraviolet light results in surface modification of polymeric materials due mostly to photooxidation, photografting and/or photodegradation. Depending on the wavelength used one may anticipate that the more penetrating (wavelength higher than 200 nm) will cause degradation of the bulk structure, whereas surface and thin films dry modifications will require a wavelength that, because of the high-absorption cross-section of the polymer, penetrates to a smaller depth. This light (wavelength between 180 and 200 nm) is also characterized by the energy of the photons exceeding that of most covalent bonds and it is reasonable to expect a high probability of inducing bond breaking by irradiation.

The pulsed excimer lasers allow one to operate in the far-UV, e.g. 13 nm (ArF laser) and at high light intensity (> 10 MW/cm^2). The added advantage of using laser light derives from the relatively high fluences reachable and from the possibility of triggering multiphoton (non-linear) processes. Furthermore, as the pulse duration is typically < 20 ns, during irradiation a steep elevation of the local temperature with consequent superficial morphological modification is expected and this is at variance with the fast heat diffusion normally taking place when the intense cw sources are used to irradiate polymer surfaces, with the temperature remaining usually below 50 C.

In keeping with this observation is the reported superficial oxidation obtained with lamp irradiation, whereas ablative photodecomposition (without concomitant oxidation) and product expulsion at supersonic velocity is observed following irradiating with lasers of proper wavelength. In the latter case, the question as to whether the ablation proceeds from photochemical decomposition of the initially formed electronically excited states or is a normal "thermal process" originating from hot ground states, has not, as yet, found a clear answer.

Srinivasan et al and Yeh have already reported on the ablative photodecomposition of polymers such as polyethyleneterephtalate (PET), polymethylmethacrylate (PMMA) and polyimide (PI) (1-4), we have studied the surface modifications that polymers such as polypropylene (PP), polyethylene (PE), polytetrafluoroethylene (PTFE), polycarbonate (PC) and polystyrene (PS) undergo when irradiated with UV excimer lasers

155

F. Garbassi and E. Occhiello (eds.), High Energy Density Technologies in Materials Science, 155–159.
© 1990 Kluwer Academic Publishers.

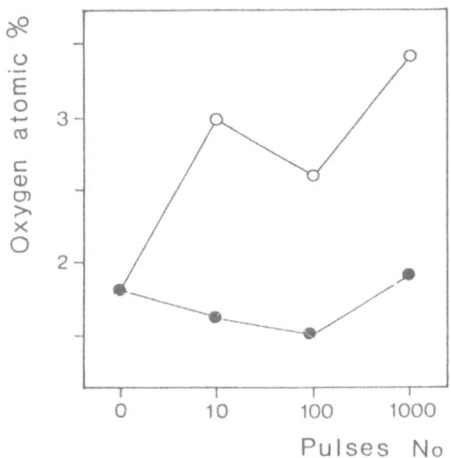

Fig. 1. XPS oxygen concentration (atomic %) for PP treated with 193 nm (o) and 248 nm (●) laser irradiation as a function of number of pulses.

at wavelengths ranging from 193 to 350 nm and of polyetheretherketone (PEEK) photolyzed with UV and infrared lasers.

PP, PE, PTFE

A threshold fluence of 80 mJ/cm² was needed in order to observe ablative photodecomposition, whereas increasing the number of pulses resulted in a progressive change of the surface from initially rough, to smooth, to structured polycrystalline phase (5).

The XPS analysis of the photochemically modified polymer revealed a consistent increase in the oxygen/carbon ratio (Fig. 1), indicating that the intermediate carbon-centered radicals were scavenged by molecular oxygen to give alcohol or ketone functional groups via the hydroperoxy radicals. We have found no differences in the surface modification on going from the 193 to the 248 nm laser light.

PS and PC

The laser treatment of polystyrene (PS) and polycarbonate (PC) was performed at the same wavelengths (6). In the case of PS the etching is evident only at 193 nm, whereas the longer

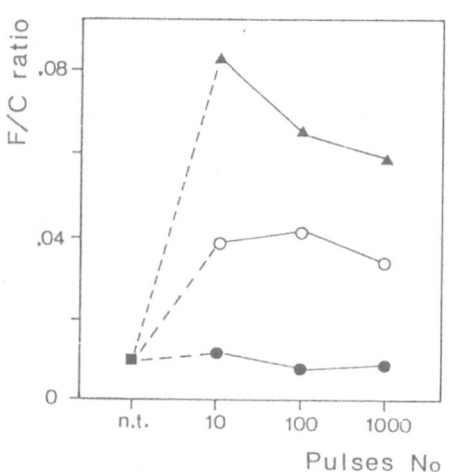

Fig. 2. XPS F/C ratio for PC treated with 193 nm (O), 248 nm (▲)
and 350 nm (●) laser irradiation as a function
of number of pulses.

wavelengths (248 and 350 nm) induced surface modification due
essentially to thermal effect and subsequent thermal shock.
Furthermore no photooxidation of the surface has been evidenced.

The polycarbonate behaves differently, as expected on
structural grounds, and the laser irradiation (248 nm) leads to
oxygen depletion and photo-Fries rearrangements and at 193 nm to
oxygen depletion and etching. The rearrangements are easily
evidenced by "tagging" the resulting functionalities (usually -OH
and -COOH groups) with atoms of high photoemission cross-section
as e.g. the fluorine (4.5 times more intense than carbon) and
exploiting the following reactions:

$$-OH \ + \ -(CF_3CO)_2O \ ----> \ -OOCCF_3$$

$$-COOH + \ -(CF_3CO)_2O \ ----> \ -OCOOCCF_3$$

which easily occurs by exposing the treated polymer to the
vapours of trifluoroacetic anhydride (Fig. 2).

Photolisis at 350 nm (X = F) does not induce any chemical
modification on the surface of either PC or PS, and the
microcrazing observed (PS) may be due solely to thermal effect,
indicating a degradation of the photochemical energy into heat.

Quite peculiar are the cone-like structures originating from
microparticulates that shield the underlying polymer from laser
radiation during photolysis. It is also interesting to note that
from fluence/type of modification studies we have no evidence for
the occurrence of multiphoton (non-linear) effects.

158

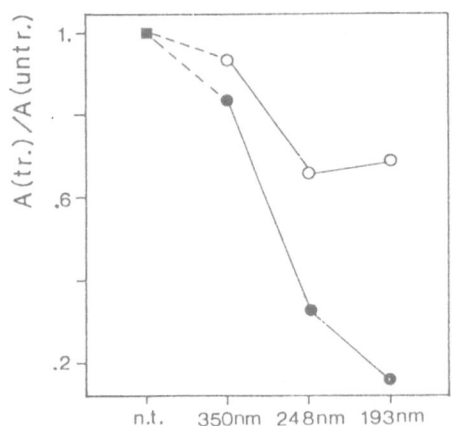

Fig. 3   Plot of the A(tr.)/A(untr.) ratio, A is the percentage
         of the XPS C-1s spectrum relative to a particular
         component: C-O (o), C=O (●).

PEEK

The polyetheretherketone (PEEK) are interesting materials
and seem to be attractive as high quality engineering
thermoplastic and the repetitive unit is made up of phenyl rings
linked by two ether and one ketone bridges. Thermal properties of
PEEK are very similar to those of PET.

The treatment with infrared and excimer lasers operating at
high fluence values (2-5 J/cm²) leads to surface modification,
due, essentially, to thermal effects. Furthermore etching was
induced by focussed beams and the morphology of the treated
surface was essentially smooth and independent of the photolyzing
wavelength (193, 248, 350 nm). By contrast, a wavelength
dependence was observed when operating at low fluences (< 100
mJ/cm²) (7).

As in the case of PC the etching at 193 nm left behind a
rough surface with "cone-like" structure domains. At 248 nm (55
mJ/cm²) the surface structure was more irregular because of the
redeposition of oligomers in the form of debris. The longer
wavelength radiation, i.e. 350 nm, did not induce any noticeable
modification.

The XPS studies revealed a net decrease in the O/C ratio,
which was more substantial at 193 nm and tapered off at 350 nm,
consistent with the effectiveness of the photochemically induced
modification at shorter wavelength (Fig. 3). Tagging experiments
with TFFA were performed and the results ruled out the formation,
by photolysis, of hydroxyl and/or carboxyl groups as were instead
observed in the case of polycarbonate. We suggest that the oxygen

containing functionalities are present on the volatile compounds. The CO_2 laser treatments did not result in any significant chemical modification of the polymer surface, beside a physical reorganization due to melting-solidification cycles.

We can, at this point, conclude that the incidence of photochemical and thermal processes induced by laser photolysis depends on the fluence. At high fluence the thermal effects dominate, whereas the chemical modification (more efficient elimination of C=O) following photochemical events is important at lower fluence values.

References

1. R. Srinivasan, B. Braren, J. Polym. Sci. Chem. Ed., 22, 2601 (1984)

2. R. Srinivasan, B. Braren, D. E. Seeger, R. W. Dreyfus, Macromolecules, 19, 916 (1986)

3. V. Srinivasan, M. A. Smrtic, S. V. Tabu, J. Appl. Phys., 59, 3861 (1986)

4. J. T. C. Yeh, J. Vac. Sci. Technol, B4, 653 (1986)

5. V. Malatesta, F. Garbassi, E. Occhiello, "Proceedings of the International Conference on Lasers '86", p. 179 (1987)

6. E. Occhiello, F. Garbassi, V. Malatesta, J. Mat. Sci., 24, 569 (1989)

7. E. Occhiello, F. Garbassi, V. Malatesta, Angew. Makromol. Chem., 169, 143 (1989)

WETTING BEHAVIOR OF OXYGEN PLASMA TREATED PTFE

M. Morra, E. Occhiello, F. Garbassi

Istituto Guido Donegani S.p.A.
Via Fauser 4
28100 Novara ITALY

Abstract

PTFE was treated with oxygen plasma and the effects of treatment time were evaluated by XPS, SSIMS and contact angle measurements. The behavior of water and CH_2I_2 advancing and receding angles was interpreted on the basis of current theories on contact angle hysteresis. At short treatment time wettability reflects chemical modification of the surface, while at longer treatment times surfaces are deeply etched and contact angles are controlled by roughness. The typical behavior of composite surfaces was observed when using water as the wetting liquid.

Introduction

Plasma treatment frequently used to modify the wetting behavior of polymers (1-3). The purpose of this communication is to describe the wetting behavior of oxygen plasma treated PTFE as a function of treatment time using contact angle hysteresis, XPS and SSIMS to account for alterations of surface energy and chemistry, respectively. Former studies on plasma treated PTFE surfaces (4-6) rarely discussed contact angle hysteresis. In a preliminary communication we described the morphological alterations occurring during etching (7).

Experimental

Plasma etching experiments were performed using a parallel plate reactor, with the samples located on the water-cooled grounded electrode. The plasma parameters were the following: excitation frequency 13.56 MHz, power 100 W, pressure 2 Pa, gas flow 8 cc(STP)/min. Oxygen from lecture bottles supplied by Carlo Erba was used. PTFE plaques (1 mm thick) were kindly provided by Montefluos.
Contact angles were obtained by the sessile drop technique, on a Rame'-Hart contact angle goniometer. Advancing and receding angles were obtained by increasing or decreasing the drop volume until the three phase boundary moves over the surface (1). The capillary pipette of the microsyringe was kept immersed in the drop during the entire measurement, as described in (1). Reported values are averaged over at least 10 different measurements, performed in different parts of the specimen surface.

F. Garbassi and E. Occhiello (eds.), High Energy Density Technologies in Materials Science, 161–168.
© 1990 *Kluwer Academic Publishers.*

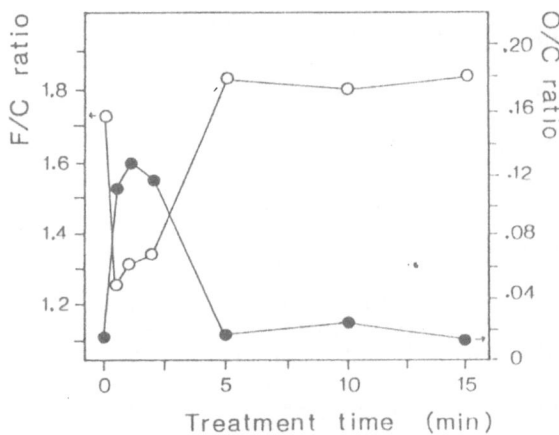

Fig. 1.

XPS O/C (●) and
F/C (o) ratios
as a function of
treatment time.

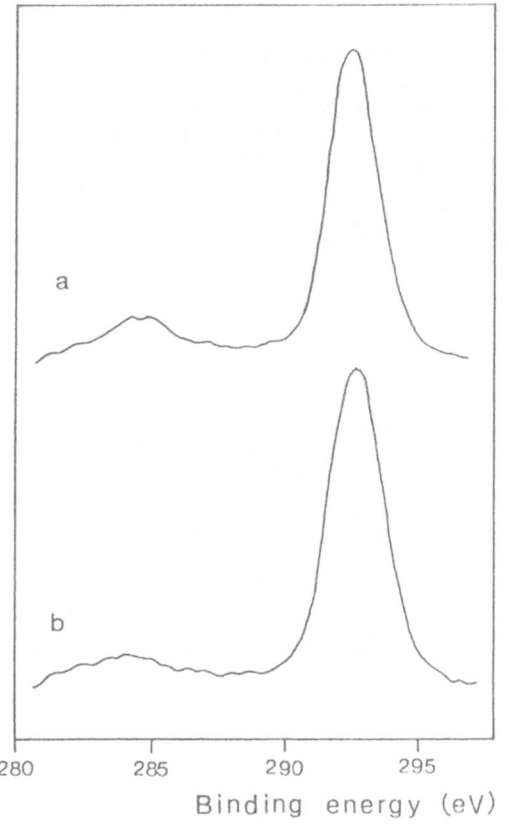

Fig. 2.

XPS C1s peak of
untreated PTFE (a)
and of the 15 mins.
treated sample (b).

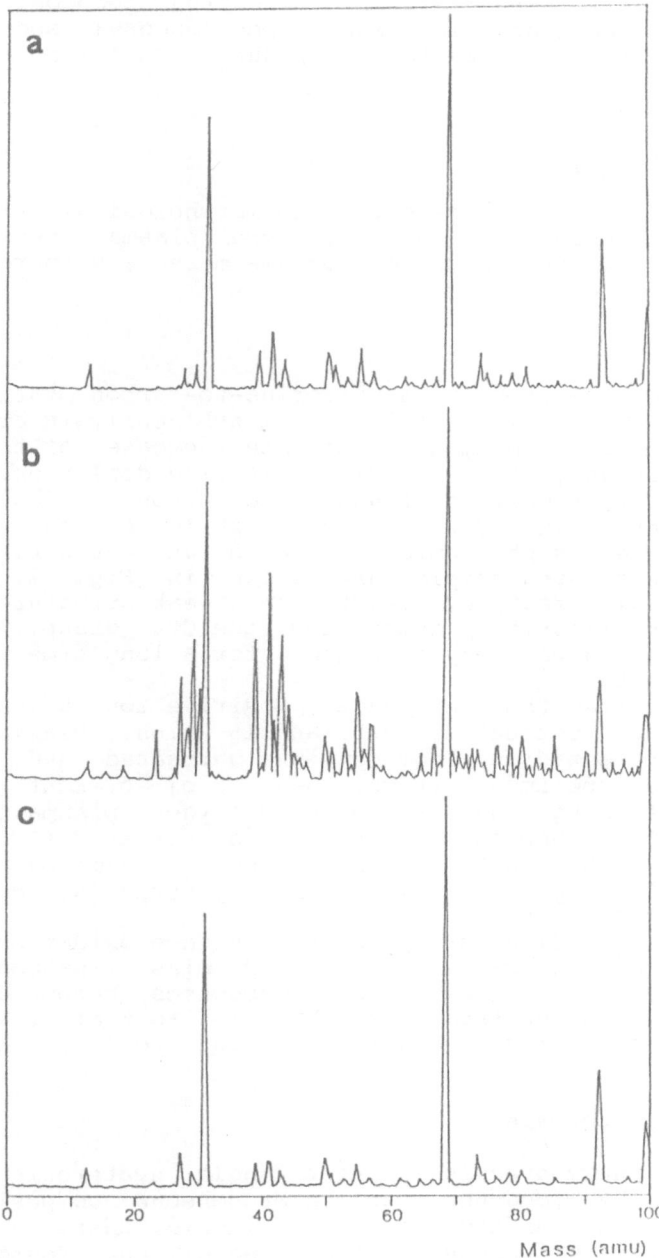

Fig. 3.   SSIMS positive ions spectra relative to a) untreated
          sample b) after 0.5 mins. c) 15 mins.

Carefully deionized and bidistilled water and analytical grade freshly distilled CH_2I_2 (Carlo Erba) were used for contact angle measurements. XPS (X-ray Photoelectron Spectroscopy) and SSIMS (Static Secondary Ion Mass Spectroscopy) spectra were obtained with the experimental procedures described elsewhere (8).

Results and discussion

SEM micrographs are showed that no morphological evidence of etching appeared below 2 minutes of oxygen plasma treatment. At higher treatment times the surface became more and more etched, finally spongy-like (7).

XPS and SSIMS

The oxygen-to-carbon and fluorine-to-carbon ratios as a function of treatment time obtained from XPS analysis are shown in Fig. 1. At low treatment time some increase of the oxygen content was observed, together with fluorine depletion. Further oxygen plasma treatment led fluorine-to-carbon ratios close to the theoretical value (F/C = 2), while oxygen virtually disappeared. The C1s photoemission peak of untreated PTFE and of the 15 minutes treated sample are shown in Fig. 2. In both spectra, the only feature, apart from a weak structure due to hydrocarbon contamination, comes from the CF_2 group. Therefore the samples treated with oxygen plasmas for a long time look like a very clean PTFE.

SSIMS supported this evidence. Positive ion SSIMS spectra corresponding to untreated, 0.5 and 15 mins., plasma treated samples are presented in Fig. 3. The untreated and 15 mins. plasma treated specimens (Figs. 3-a, 3-c) present the same spectrum, confirming that a prolonged oxygen plasma treatment produces a surface chemically identical to untreated PTFE. On the other hand the 0.5 mins. treated samples present hydrocarbon-related peaks in the SSIMS spectrum (8), confirming fluorine depletion.

Our XPS and SSIMS data point to surface oxidation in the initial stages of the treatment (up to 2 mins. treatment time). At higher treatment time etching processes become dominant, the chemistry of the surface becomes similar to that of untreated PTFE but extensive erosion occurs, leading to a rough surface (7).

Contact angle hysteresis

True or thermodynamic contact angle hysteresis (3) is accounted for by roughness or heterogeneous composition (1, 9-10). For an ideal surface and a given liquid, there is only one thermodynamically definable contact angle, the Young contact angle. The transition to a non ideal surface results in a change of the equilibrium contact angle (i.e. the absolute minimum of the free-energy versus contact angle plot) and in the observation of a range of contact angles corresponding to different allowed metastable states.

The equilibrium contact angles for rough and heterogeneous surfaces are described by the Wenzel (11) and Cassie equations (12), respectively. The Wenzel equation (1) takes into account the effect of increased area, while the Cassie equation (2) states that the equilibrium contact angle on a two-phase heterogeneous surface is the weighted average of the corresponding contact angles.

$$\cos\theta_w = r \cos\theta_y \qquad\qquad\qquad (1)$$

$$\cos\theta_c = Q_1 \cos\theta_1 + Q_2 \cos\theta_2 \qquad\qquad (2)$$

In eq. (1) r is the ratio between effective area and geometric area (roughness factor), θ_w is the Wenzel angle (the contact angle corresponding to the absolute minimum of the free energy on the surface with roughness factor r) and θ_y is the Young angle (the stable equilibrium contact angle on the corresponding smooth surface). Since r is always bigger than unity the effect of roughness is to increase (if $\theta_y > 90°$) or to decrease (if $\theta_y < 90°$) the equilibrium contact angle. In eq. (2) θ_c is the Cassie angle (the equilibrium contact angle for the liquid drop on the heterogeneous surface), while θ_1 and Q_1 are the Young angle and the fraction of surface relative to phase i respectively.

The departure of the surface from ideality leads to the formation of many closely spaced metastable states (local minima). Therefore the system exhibits a range of allowed contact angle values. Since it is impossible to distinguish the true minimum from the other minima, θ_w and θ_c cannot be measured. However, rough and heterogeneous surfaces can be conveniently characterized by the measurement of the highest possible value, the advancing angle, and the lowest possible value, the receding angle, their difference being the contact angle hysteresis. Contact angle hysteresis mainly depends, among other factors, on the height of the energy barriers between metastable states. Calculations on model surfaces, showed that in general hysteresis increases with roughness. However, surfaces having an intrinsic contact angle greater than 90° may show a dramatic decrease of hysteresis at high roughness. This behavior reflects the transition to a "composite" surface, characterized by so huge a roughness that a liquid with high intrinsic angle cannot completely wet the crevices. In this case the surface beneath the liquid drop is either solid or air, hence the name "composite". The equilibrium contact angle for a composite surface is described by the Cassie and Baxter equation (13) (actually a special case of equation (2) with the contact angle of the liquid on air equal to 180°). In eq. (3) θb is the equilibrium or Cassie and Baxter angle, θ is the Young contact angle on the corresponding smooth surface and Q_1 and Q_2 are the fractions of wetted and unwetted region.

Fig. 4.

Advancing (o)
and receding (●)
H_2O contact angles
as a function
of treatment time

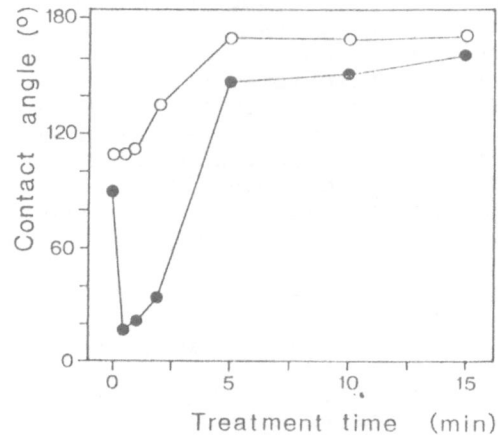

Fig. 5.

Advancing (○)
and receding (●)
CH_2I_2 contact angles
as a function
of treatment time

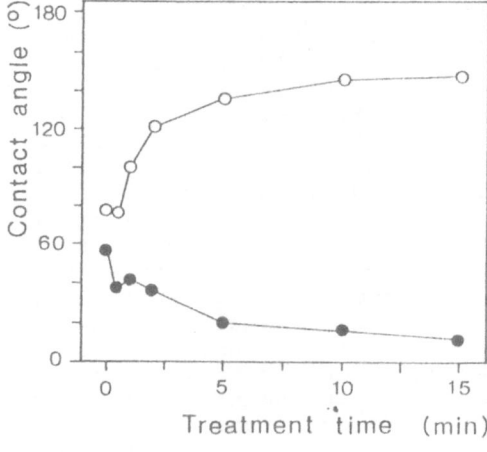

$$\cos\theta_b = Q_1 \cos\theta - Q_2 \qquad\qquad (3)$$

On composite surfaces the height of the energy barriers between metastable states decreases dramatically and contact angle hysteresis is greatly lowered. Theoretical results have been confirmed experimentally, at least qualitatively (2). Also a heterogeneous composition greatly affects the observed angles. Taking into account a model heterogeneous surface composed by a high energy and a low energy part, the fraction of surface covered by the high energy component must be close to unity, in order to decrease the advancing angle. On the contrary a small amount of high energy component is sufficient to cause a drop in the receding angle. Thus the advancing angle is characteristic of the low energy portion of the surface while the receding angle reflects the high energy portion.

On the basis of the contact angle hysteresis theory briefly outlined above, the wetting behavior of oxygen plasma treated PTFE can be understood. Fig. 4 shows the advancing and the receding angle of water on treated PTFE as a function of treatment time. The advancing angle of water on the untreated sample is very close to the tabulated Young angle (108° (1)). The receding angle is about 20° lower, probably because of the small amount of oxygen observed by XPS on the untreated surface. After 0.5 mins. treatment, the oxygen content increases: the presence on the surface of a high energy component is recorded by the decrease of the receding angle; however the oxygen·amount on the surface is not enough to affect also the advancing angle. The latter begins to change after 2 mins. treatment. The increased value of the advancing angle together with the SEM evidence indicate that roughness contributes to wettability. After 5 mins. treatment, surfaces become absolutely hydrophobic. Water drops roll very easily on the surface due to the high advancing angle and low hysteresis. In fact, due to the absence of low value receding angles, no obstacle to their movement occurs, as characteristic of a composite surface. SEM showed that at this stage the surface is deeply etched so water cannot penetrate cracks and crevices (7).

The measurement of CH_2I_2 contact angles provided important supplementary information. Since the Young contact angle of CH_2I_2 on PTFE is lower than 90°, according to theory no transition to a composite surface is possible. Thus a monotonic increase of hysteresis with roughness must be expected. CH_2I_2 contact angles reported in Fig. 6, are in full agreement with theoretical expectations.

Conclusion

The wetting behavior of oxygen plasma treated PTFE reflects, at short treatment time, the modification of surface chemistry due to the introduction of some oxygenated groups and fluorine depletion. At longer treatment times the surface chemically identical to PTFE but deeply etched, therefore wettability is controlled by roughness.

Acknowledgments

We thank Mr. L. Pozzi for XPS spectra and Mr. G. Morelli for SEM micrographs.

References

1. S. Wu, "Polymer Interface and Adhesion"; Marcel Dekker, New York, 1982

2. R. E. Johnson, Jr., R. H. Dettre, in "Surface and Colloid Science", E. Matijevic Ed.; Wiley-Interscience, New York, 1969, Vol. 2.

3. J. D. Andrade, L. M. Smith, D. E. Gregonis, in "Surface and Interfacial Aspects of Biomedical Polymers", J. D. Andrade Ed., Plenum Press, New York, 1985, Vol. 1, Chapter 7.

4. H. Yasuda, H. C. Marsh, S. Brandt, C. N. Reilley, J. Polym. Sci. Chem. Ed., 15, 991 (1977)

5. R. H. Hansen, H. J. Schonhorn, J. Polym. Sci., B4, 203 (1966).

6. J. R. Hollahan, B. B. Stafford, R. D. Falb, S. T. Payne, J. Appl. Polym. Sci., 13, 807 (1969).

7. M. Morra, E. Occhiello, F. Garbassi, Langmuir, 5, 872 (1989)

8. F. Garbassi, E. Occhiello, F. Polato, J. Mater. Sci., 22, 207 (1987); ibid. 22, 1450 (1987).

9. R. E. Johnson, Jr., R. H. Dettre, J. Phys. Chem. 68, 1744 (1963).

10. A. W. Neumann, R. J. Good, J. Coll. Interf. Sci., 38, 341 (1972).

11. R. N. Wenzel, Ind. Eng. Chem., 28, 988 (1936).

12. A. B. D. Cassie, Discuss. Faraday Soc., 3, 11 (1948).

13. A. B. D. Cassie, S. Baxter, Trans. Faraday Soc., 40, 546 (1944).

AUTHOR INDEX

170